CYBER SECURITY CHALLENGES AND CYBER SECURITY STRATEGIES IN SOME EUROPEAN (VISEGRAD) COUNTRIES AND ADAPTATION FOR ASIAN COUNTRIES: THE CASE OF VIETNAM

Zoltan Rajnai – Nguyen Huu Phuoc Dai

Cyber Security Challenges and Cyber Security Strategies in Some European (Visegrad) Countries and adaptation for Asian countries: the case of Vietnam

L'Harmattan
Paris
2022

EST. 1991

Battyány Lajos
FOUNDATION

L'Harmattan France
5-7 rue de l'Ecole Polytechnique
75005 Paris

ISBN 978-2-14-026529-7
DOI: https://doi.org/10.14324/978-2-14-026529-7

http:/www.harmattan.fr
diffusion.harmattan@wanadoo.fr
www.amazon.com

Cover design: Zsolt Gembela
Page-setting: Krisztina Csernák

TABLE OF CONTENT

INTRODUCTION

In 2019, according to global risks report, there are five major threats such as extreme weather events, failure of climate change, mitigation and adaptation, natural disasters, data fraud or theft, and cyber-attacks. It can be seen that two main global threats related to information loss or damage from computer attacks in the cyberspace are mostly security concerns in many countries. Currently, cyber crimes are more complicated in their attacks due to the development of technology. There are various types of cyber crimes such as malware, spyware, virus, ransomware, and the like. They can cause serious impacts not only individuals but also organizations, governments or nations. Cyber crimes have no border between one nation and the others because it mainly depends on the Internet environment; as a consequence, the prevention or the fighting against the cyber-attacks are not the responsibility of one country but whole countries in the world. Hence, a legal framework of cybersecurity cooperation, cybersecurity strategy or general model of cybersecurity cooperation is an essential and urgent thing in the modern technology era. To achieve this, this research aims to figure out the answers for several essential questions, as follows:

1. What are the current cyber-threats?
2. Are they dangerous threats to Vietnam?
3. What can Viet Nam do to mitigate cyber-threats?
4. How can Viet Nam cooperate with international organizations to solve these risks?
5. What are the benefits of Visegrád strategies in defense towards cyber-attacks with other countries?
6. Which cybersecurity strategies are found suitable for Vietnam in terms of adaptation?

The thesis of this reseach can bring a general overview of the current framework of cybersecurity cooperation among Asia, ASEAN, and EU nations. Moreover, this thesis may clarify the difference of strategies, structures or purposes in cybersecurity cooperation in these areas. Remarkably, the author points out the special case in cybersecurity cooperation of Visegrád countries on the way to protect themselves and their contributions to the other nations in the same region towards global cyber-threats. Furthermore, the author also proves that some cybersecurity strategies may be applied in ASEAN countries, particularly in Vietnam. In addition, the researcher also offers some solutions in cooperation amongst ASEAN countries, especially in Vietnam and its neighbors. The most successful thing of the thesis is that it reaches the aims of the research when it offers clear answers for assumptions, as follows:

1. Cybersecurity in Visegrád countries shares similarities in goals, strategies, and strength to align with European Union Member States regarding armed forces, cybersecurity, and national security.
2. Cybersecurity in the East Asian and the South East Asian countries aim to create a more secure society and supports economic development.
3. Singapore's cybersecurity strategy may be adapted to Vietnam's legal framework.
4. Cybersecurity, especially in cybersecurity cooperation in Visegrád countries may be adapted and networked with Asian countries, particularly in Vietnam and its neighbors.

INTRODUCTION

A. THE THESIS

Nowadays, computers and the Internet are becoming increasingly indispensable tools in several aspects of our lives including academic study, professional work, entertainment, and communication. In particular, the booming of the Internet of Things (IoT) benefits not only individuals but also businesses and society. Firstly, this concept refers to safety, health, and finance for an individual by tracking the health signs of victims and providing appropriate medical treatments. Moreover, IoT can allow people to monitor the security of their houses through mobile or smart devices. Secondly, IoT can help companies or factories to improve operations and increase customer satisfaction by tracking their goods during shipping, location, control, and security[1]. Although there are a lot of benefits accompany the Internet or IoT and cyberspace, they are full of vulnerabilities, threats, and security issues [2]. Therefore, computer and network security are a concern not only for traditional security awareness organizations; for example, military, bank, or financial institutions but also for every individual and government officials who use computers. Besides, nowadays more and more organization's valuable assets are stored in the computerized information system; the security of the system has become an essential and urgent issue [3]. However, it is remarkable that most of the systems today are designed with little attention to security concerns. Viet Nam, a small country in the Southeast of Asia, is also one of the top 20 populous countries in the world but Vietnam's Information and communication technology (ICT) industry has just developed recently. In fact, in 1997, Viet Nam began to connect the world via the Internet. Despite the late start, Viet Nam approached rapidly modern telecommunications infrastructure in the world such as high-speed fiber optic cable system, VOIP, ADSL, WLAN, and WiMAX. Nevertheless, there are a lot of vulnerabilities in the current ICT network system and the management of a network system is quite weak. At the beginning of approaching ICT, Viet Nam government did not pay attention to the network security or the damage from the cyber-threats. This is not a question about the safety of the websites, the applications, the Internet users or the government computer's system. Rather, it is about the safety of the Vietnamese people. For example, in 2014, there were more than 200 websites were attacked by Chinese hackers including six government agencies' websites that have "gov.vn" domain [4] (a report from BKAV - a famous IT and network security company in Viet Nam). Moreover, Kaspersky Lab noted that the percentage of industrial computers was attacked from 17% in July 2016 to more than 24% in December 2016. Among them, Viet Nam is the top of three targeted-attack countries with more than 66%, Algeria (over 65%) and Morocco (60%) [5]. Furthermore, the dangerous attack occurred on 29th of July 2016, the official website of Viet Nam Airlines was hijacked at some international airports as Noi Bai, Tan Son Nhat, Da Nang, and Phu Quoc by a Trojan named (Trojan.Win32.Dropper.Encrypt.K.). The users were redirected to another website that contained false information. It led to 400,000 Golden Lotus member's data which were published on the website such as name, birthday, workplace, address, nationality, telephone number, password, and so on [6]. Then, the perpetrator was identified by a Chinese hacker group named 1937CN – the strongest hacker group in China. Moreover, this group also attacked around 1000 Vietnamese websites among 15 government websites with the domain (gov.vn), 50 education websites (edu.vn) and around 200 websites of the Philippines on the last two days of May in 2015 [7]. Therefore, if the Vietnamese critical infrastructure is threatened or damaged, it will lead to unimaginable effects not only for the government but also for Vietnamese citizens. These damages influenced Vietnamese critical infrastructure, especially in air transportation.

In order to investigate whether Viet Nam is ready to face any security problems or find out solutions for them, this study seeks the answers to the six following important questions:

1. What are the current cyber-threats?
2. Are they dangerous threats to Vietnam?
3. What can Viet Nam do to mitigate cyber-threats?
4. How can Viet Nam cooperate with international organizations to solve these risks?
5. What are the benefits of Visegrád strategies in defense towards cyber-attacks with other countries?
6. Which cybersecurity strategies are found suitable for Vietnam in terms of adaptation?

Besides, the scope of this thesis mainly focused on the cybersecurity strategies of Visegrád countries and some neighbor countries of Vietnam in the ASEAN area in order to propose the new cybersecurity framework for Vietnam and its' neighbor countries. Threfore, the most successful thing of the thesis is that it reaches the aims of the research when it offers clear answers for assumptions, as follows:

1. Cybersecurity in Visegrád countries shares similarities in goals, strategies, and strength to align with European Union Member States re-

garding armed forces, cybersecurity, and national security.
2. Cybersecurity in the East Asian and the South East Asian countries aim to create a more secure society and supports economic development.
3. Singapore's cybersecurity strategy may be adapted to Vietnam's legal framework.
4. Cybersecurity, especially in cybersecurity cooperation in Visegrád countries may be adapted and networked with Asian countries, particularly in Vietnam and its neighbors.

B. BOOK OUTLINE

The structure of the book is described in four chapters, as follows:

Chapter One reviews the literature on cyber threats, cybercrimes, cyber attacks, and the like are overviewed. Moreover, it clarified the differences between cyber-crime and cyberwar in order to take into account the new trends of cyber security and cyber threats. Furthermore, it primarily expressed an urgent need for Vietnam's cybersecurity strategies toward the new cybersecurity trends in the world.

Chapter Two describes cybersecurity, policies, strategies, cooperation in Visegrád countries, and ways to do those things. Besides, it also points out cybersecurity cooperation and legal frameworks in the EU like ENISA, NATO, the Three Seas Initiative, Digital Single Market Initiative, NIS directive, GDPR, NIST 800-53, Contractual Public-Private Partnership (CPPP), and European Public-Private Partnership for Resilience (E3PR).

Chapter Three presents the policies and strategies of each Asia and ASEAN countries; and cooperation among these countries. Besides, it expresses the different major factor between Asia and EU nations is data protection regulations. In addition, it shows several strong and weak countries about cybersecurity capacity building in Asia and ASEAN.

Chapter Four presents suggestions from Visegrad countries' cybersecurity strategies that can be applied in Asian countries, including Vietnam. Some certain initiatives for Vietnam and its neighbors are also recommended to enhance their position in the global integration era as a group of countries like the Visegrad Group's countries.

Chapter One:

CYBERSECURITY IN THE EAST ASIAN AND SOUTH EAST ASIAN COUNTRIES

Currently, there is a growing concern about cyber threats, the most dangerous ones all over the world. They can cause huge damage to finance, the economy, politics, and other aspects of life. As a result, identifying types of cyber threats is a critical thing and an urgent need not only for individuals and businesses but also for governments and organizations to increase awareness of cybersecurity, national security and find solutions to mitigate or reduce the damage from them. Moreover, it was expected to figure out the differences in security cooperation among Asian nations in order to identify which model is suitable for small nations, including Vietnam and its neighbors in the Asian region. Regarding these purposes, a review of literature relevant to this study was considered from scientific articles, trustworthy sources from experts, and the Internet in order to provide an overview of types of cyber threats, their impacts, and security cooperation between Asian countries with others.

1.1. THEORETICAL BACKGROUND

Threats

Cybercrimes are used by many complex techniques to encode some complicated algorithms in malware, spyware, and virus in order to put them on the Internet and take advantage of breaching them. Moreover, cyber-attacks can cause a dangerous impact to an organization such as economic impact [8] [9], reputation, loss of sensitive business information, lack of trust, business disruption, equipment loss, and stock prices [10]. Most of the attacks cause bad consequences in the financial problems of an organization. Moreover, when a firm is under attack, it faces losing the trust of its customers and people are afraid to invest more in this company. Furthermore, due to DoS/DDoS attacks, they can stop the services of the company, which leads clients to move to other services of other competitors. In addition, in some cases, malware can destroy whole network equipment (like the Stuxnet worm) [11], the company needs to spend a lot of money reinstalling the system. In May 2017, there is a dangerous attack on many countries in the world from a new malware - named

Wanna-cry ransomware [Figure 1.1]. This ransomware is available in 28 different languages [12]. This malware seems an obsession horror for every country under attack. The victim needs to pay a lot of money to take back data; unless it spreads out too fast and destroys all data by a countdown timer.

Figure 1.1: The countries are attacked by WannaCry ransomware [12]

With the rapid development of multimedia and networking, it offers some benefits for the users; however, they also face many kinds of hazardous threats that can cause serious problems.

Transnational crimes

Transnational organized crime (TOC) is defined as the nonmilitary threats that shrink the economic, political, and social aspects of a nation or the citizens' health [13]. Transnational threats have many categories such as arms trafficking, drug trade[14], human trafficking, the weapon of mass destruction (WMD) [15], people smuggling, terrorism, and financial crime[16]. In another hand, transnational crimes are also known as an international organized crime that targets government agencies. For example, in 2009, the United Nations office on drugs and crime (UNODC) said that drug trafficking brought huge profits every year (about 68 billion dollars of global cocaine and 85 billion dollars opiate markets alone) [17]. Moreover, International Labor Organization (ILO) reported that in 2005 the number of victims is used for sexual or labor-based exploitation (men, women, and children) approximately 2.4 million people with annual profits of 32 billion dollars. In particular, in Europe, human trafficking for sexual exploitation brings 3 billion dollars with 140.000 victims every year [17]. In addition, regarding the report of transnational crime and the developing world in 2017, there are 11 types of transnational crimes which increased the profit between 1.6 trillion USD and 2.2 trillion USD per year [18],[Table 1.1]. As a consequence, transnational crime may cause financial violence and corruption, damage the environment and citizen's life in the world.

Table 1.1: The Retail value of Transnational crime (Source: [18])

Transnational Crime	Estimated Annual Value (USD)
Drug trafficking	$426 billion to $652 billion
Small Arms & Light weapons trafficking	$1.7 billion to $3.5 billion
Human Trafficking	$150.2 billion
Organ Trafficking	$840 million to $1.7 billion
Trafficking in Cultural Property	$1.2 billion to $1.6 billion
Counterfeiting	$923 billion to $1.13 trillion
Illegal Wildlife Trade	$5 billion to $23 billion
IUU Fishing	$15.5 billion to $36.4 billion
Illegal Logging	$52 billion to $157 billion
Illegal Mining	$12 billion to $48 billion
Crude Oil theft	$5.2 billion to $11.9 billion
Total	$1.6 trillion to $2.2 trillion

Transnational organized crime groups have dramatic threats not only in one country but also in the global region. They can cooperate with local criminals in order to increase disruption, extortion, gangster, and violence. They may damage countries' stabilization and put citizens' lives at risk.

Cooperation in the cybersecurity of Asian countries

Several security challenges of Asian countries include nuclear proliferation, terrorism, cross-border crime, pandemics, natural catastrophes, resource conflicts, major power rivalries, piracy, and the like [19]. Moreover, there are several forums that overlap cooperation with each other in East Asia like Association of Southeast Asian Nations (ASEAN), ASEAN Regional Forum (ARF), Asian-Pacific Economic Cooperation (APEC), East Asia Summit (EAS), ASEAN Defense Ministers Meeting (ADMM), Council for Security Cooperation in the Asia Pacific (CSCAP) and Expanded ASEAN Maritime Forum (EAMF) [Figure 1.2]. China is considered as a leading country of Asia-Pacific with the fast growth in economy and military. Hence, it led China as a threat to other countries in the same region. However, there are four bilateral cooperation between China and Asia countries like Sino-India, Russia-China, USA-China, and North Korea- China in economic, diplomatic, and military [APPENDIX 2]. On the other hand, the USA is also a traditional and strong power country in the military in the world, as a result, the main concern of ASEAN countries is that staying away from choosing the cooperation between the USA and China. In 2003, the ASEAN community wanted to achieve cooperation in 3 major pillars: security, economic, and socio-cultural cooperation by 2020 [20]. Moreover, there is five bilateral cooperation between

Asian countries with the US such as US-Japan, US-South Korea, US-Australia, US-Philippines, and US-Thailand in security platform [APPENDIX 1].

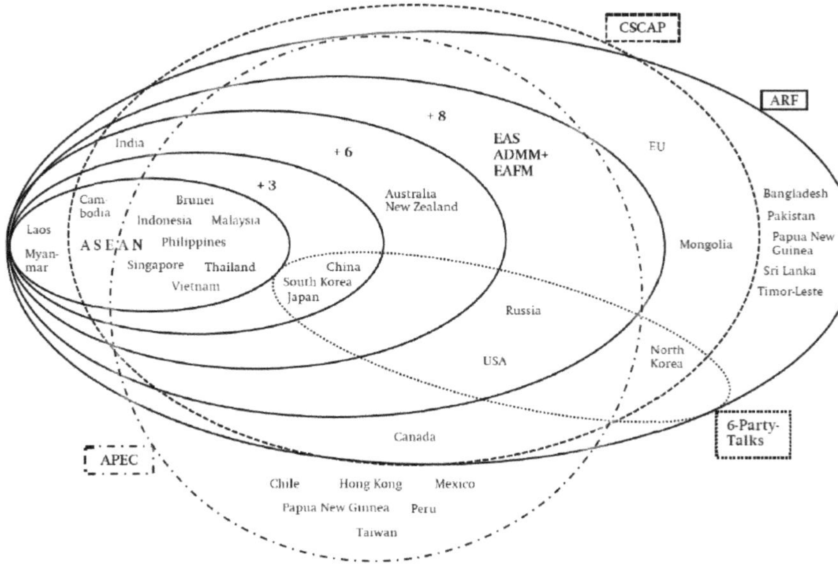

Figure 1.2: Regional Formats in East Asia and their overlaps (Source: [19])

USA-Japan cooperation

The cooperation between the USA and Japan started in the late 18th century and early in the 19th century. This bilateral cooperation involved military, economy, science, technology, and politics for the information and technology exchange. Japan, one of the closest allies and partners of the USA, supported the USA in missile defense system development. Moreover, Japan established an alliance coordination mechanism and expand in maritime security, cyberspace, and outer space for the USA. For instance, in September 2011, the first meeting of bilateral strategic policy dialogue on cybersecurity was held to share the views on security challenges in cyberspace. Moreover, they improved the security system and enhanced counter-intelligence measures via intelligence activities. Indeed, the USA – Japan alliance was remarked in 2015 via the release of the revised USA-Japan defense guidelines that expanded forms of security-oriented cooperation [21]. In 2016, they signed a new five-year package of host nation assists for the USA army in Japan. In order to promote security and defense cooperation, Japan and the USA also built trilateral security and defense cooperation with both Australia and the Republic of Korea [22].

USA-South Korea cooperation

This cooperation was founded in 1950 when the USA supported South Korea became a modern state – known as the Republic of Korea. The USA protected South Korea from North Korea and controlled the operation of South Korea's army. Actually, in 1953 South Korea and the USA agreed to a military alliance together [21]. South Korea became an important economic partner with the USA. For example, South Korea was the seventh-largest market for USA goods and the second-largest market for its agricultural products in 1989 [23]. In 2001, the USA and South Korea signed a free trade agreement in order to enhance economic trading together.

USA- Australia cooperation

The diplomatic relation between the USA and Australia began in 1940. Australia is an essential ally, partner, and friend in economics, academia, and the military with the USA [24]. In fact, the Australia – United States Free Trade Agreement (AUSFTA) was signed on 18th May 2004 and effected on 1st January 2005 [25]. Their defense relationship was first established in 1918 in the Battle of Hamel in France. Besides, Australia supported the USA in the air strikes to counter Islamic State in 2014. Furthermore, Australia helped the USA in training purposes in order to increase the number of the USA marines from 250 to 2500 people. In 2015, Australia and the USA defense agencies signed a Joint Statement on Defense Cooperation to lead for future cooperation. In 2017, Australia, the USA together with other WTO members accepted to work towards future negotiations on e-commerce [26]

USA- Philippines cooperation

The bilateral cooperation between the USA and the Philippines was established in 1951. The Philippines is also one of the oldest Asian country partners of the USA and an ally of Major non-NATO [27]. They cooperated in several aspects such as maritime security, disaster response, law enforcement, cybersecurity, and non-proliferation of weapons of mass destruction. Moreover, the Philippines gave two main military station troops (Subic Bay Naval Base and Clark Air Base) as a logistical hub and repairing or resupplying facility for the USA air force. In 2014, their relation was enhanced in defense cooperation with the Agreement on Enhanced Defense Cooperation.

USA – Thailand cooperation

USA –Thailand has established its diplomatic relations in 1818 and the Treaty of Amity and commerce in 1833 [28]. After World War II, the relationship was improved in diplomatic, security, and commercial relations. Thailand is a partner for cooperation with the organization for security and cooperation in Europe. Moreover, in 2003, it was designated as a Major Non-NATO Ally.

In 2013 they signed a historic agreement on science and technology cooperation. Moreover, Thailand and the USA have the same international organizations cooperation: United Nations, ASEAN Regional Forum, Asia-Pacific Economic Cooperation Forum, International Monetary Fund, World Bank, World Trade Organization, and Organization of American States observer.

USA – Russia cooperation

The cooperation between the USA and Russia includes diplomatic, trade, and global security cooperation [29]. This relationship peaked in the spring of 1999 under Russian President Boris Yeltsin. In 2014, this relationship became so complicated because of the crisis in Ukraine and the Syrian civil war [30]. After the Cold War, the USA and Russia worked together to avoid the development and the increase of weapons of mass destruction; counter the terrorist attacks; and enhance scientific research in space, biomedical, public health, and nuclear.

While there existed cooperation between the USA and Asian countries, some Asian countries collaborated with China [APPENDIX 2]

Russia – China cooperation

Russia – China cooperation refers to the Sino-Russian diplomatic relationship. It was founded in 1991. Their relationship had some remarkable periods such as becoming a constructive partnership, strategic partnership, friendship, and cooperation in 1992, 1996, and 2001, respectively [31]. This bilateral relationship was improved from a strategic partnership to a comprehensive strategic partnership of coordination in 2011 in order to share the same and similar positions on global issues like the UN reforms, creating new international order, and managing the global climate change [32]. In addition, their cooperation became a special relationship in military, economy, politics, energy, and culture during a state visit between Chinese President Xi Jinping and Russian President Vladimir Putin in Moscow in 2013. Furthermore, regarding the cooperation with Russia, China enhanced maritime strategy by establishing outlined in China's first military strategy in 2015 and a general plan for bilateral military collaboration for the period 2017 - 2020 [33]. Last but not least, this bilateral cybersecurity deal was made in May 2015 and it was taken into account on protection for social cyber issues.

India – China cooperation

The relationship between India and China began in 1950. It was a bilateral relationship in diplomatic and economic (namely Sino- Indian or Indo- China). This cooperation had several conflicts in the military like the Sino- Indian war of 1962, the Chola incident in 1967, and the Sino-Indian skirmish in 1987 [34]. However, both countries have successfully reset diplomatic and economic ties since the late 1980s. Indeed, in 2008, this collaboration became a trade partner and started a strategic and military relationship. In the beginning state, their bilateral relationship between them in the economic area had an exception at Free Trade Agreement because Indians were afraid of their industry not be able to compete with Chinese cheap imports. Nonetheless, when the Indian –ASEAN FTA and China – ASEAN FTA were signed respectively, an indirect agreement became effective in 2010 [35]. In 2012, the Prime minister of China and India built a goal to improve the bilateral trade between them to 100 billion USA by 2015. Currently, in the period 2017-2018, their trade achievement reached 89.6 billion USA.

USA – China cooperation

Different from the other relationship, the relationship between the USA and China is very complex [36]. They are extensive economic partnerships together. For instance, in 2018, the USA stated the largest economy in the world and China ranked the second largest one.

This bilateral relationship was a potential adversary and economic partner although they had several conflicts during the Korean and Vietnam War. Currently, their cooperation is between several areas such as economics, military, cultural, global security, defense policy, security cooperation, people to people, and sub-national areas as well as international affairs [37]. In addition, a cybersecurity agreement was signed in September 2015 between President Obama and President Xi Jinping [38] but this agreement mainly focused on economic protection. Last but not least, in October 2017 they organized an official meeting about law enforcement and cybersecurity dialogue in order to let two nations work together in enhancing global computer security and cybersecurity [37].

North Korea – China cooperation

This diplomatic relation started on 6th October 1949. In 1961 the treaty of cooperation friendship, cooperation, and mutual assistance was signed between two countries [39]. This treaty was continued twice in 1981, 2001, and its validity until 2021. They cooperated in two major aspects such as economic and energy. Moreover, during the Korean War from 1950 to 1953, China supported North Korea by sending 3 million soldiers to fight South Korea and the United Nations. In 2003, North Korea joined in a diplomatic initiative program (namely Six-Party Talks Open) with China, Japan, Russia, South Korea, and the United States [40]. Since then, the two nations have strong cooperation in security and defense issues [40].

Japan – China cooperation

Even though China and Japan are separated by the East China Sea, they have similarities in several aspects such as architecture, language, culture, religion, philosophy, and law [41]. Japan and China's relationship refer to Sino-Japanese friendship. Their cooperation and trade treaty was first signed in 1871. Then, the Sino-Japanese peace and friendship treaty; and Official development assistance (ODA) were created in 1978 and 1979, respectively. After the Second World War, their relationship became complicated because of the enmity from the history of the Japanese war, the imperialism, and maritime arguments in the East China Sea. On the other hand, the Sino-Japanese relationship is mainly based on economic and strategic rivalry[42]. In spite of the conflictions between the two nations, their collaboration has been improving and has focused on global trade Asia's economic activities, especially the trade war in 2018.

On the other hand, some countries have trilateral security cooperation both with the USA and China [Table 1.2].

Table 1.2: Trilateral security cooperation

Trilateral security cooperation	
Japan-South Korea-USA	- South Korea –USA relation began in 1950 - South Korea –USA signed an agreement about the military alliance in 1953 and became economic partners with each other. - Japan-South Korea diplomatic relation established since 1965 - South Korea and Japan are military allies of the USA - South Korea and Japan recognized North Korea as the same threat
Russia-China-USA	- Sino-Russia established in 1991 and Russia-USA in 1776 - Sino - Russia established a treaty of good neighborliness and friendly cooperation in 2001 - Russia – USA - diplomatic and trade cooperation
Japan-India-USA	- Japan-India relations began in 1949, sharing interests in maintaining the security of sea-lanes in the Asia Pacific and the Indian Ocean, military aspects such as fighting international crime, terrorism, piracy, and proliferation of weapons of mass destruction [43]. - Japan – India is a global partnership in economics [43],[44]. - India –USA bilateral cooperation in trading and investment, global security issues - India – USA is a global strategic partnership
Japan-China-USA	- Japan-China began in the mid of 19th century - Japan – USA – treaty of mutual cooperation and security - China-USA - cooperation between economies, military, and cultural.

There were some multilateral cooperation projects which overlap each other in cybersecurity and public security between Asian countries and other countries [table 1.3].

Table 1.3: Multi cybersecurity cooperation between Asia countries and other countries

Asia Pacific countries & Others	Brunei	Cambodia	Indonesia	Laos	Malaysia	Myanmar	Philippines	Singapore	Thailand	Timor-Lester	Vietnam	USA	China	Japan	South Korea	North Korea	Australia	Russia	Iran	India	Israel	Czech Republic	South Africa	Estonia	Finland	Hungary	Poland	Slovakia	Hong Kong
Brunei																													
Cambodia											o																		
Indonesia								x																					
Laos								x			o																		
Malaysia							x	x												x									
Myanmar											o																		
Philippines					x							x																	
Singapore			x	x	x									x			x			x									
Thailand												x							x		x								
Timor - Leste																													
Vietnam		o		o		o								o						x		x			x				
USA													x	x				x											
China												x		x		x		x											
Japan								x			o	x	x																
South Korea												x												x					
North Korea													x																
Australia								x				x																	
Russia												x	x										x						
Iran									x																				

Asia Pacific countries & Others	Brunei	Cambodia	Indonesia	Laos	Malaysia	Myanmar	Philippines	Singapore	Thailand	Timor-Lester	Vietnam	USA	China	Japan	South Korea	North Korea	Australia	Russia	Iran	India	Israel	Czech Republic	South Africa	Estonia	Finland	Hungary	Poland	Slovakia	Hong Kong
India					x			x			x																		
Israel									x			x		x															
Czech Republic											x																		
South Africa																		x											
Estonia															x														
Finland											x																		
Hungary																													
Poland																													
Slovakia																													
Hong Kong																													

Note: - o: public security cooperation
- x: cybersecurity cooperation

In the Asian region, although there are some countries with strong cybersecurity strategies (Russia; China; and Singapore), as can be seen in [Table 1.3], almost all Asian countries have weak cybersecurity cooperation with others in the same region in order to counter cyber threats. However, some countries have strong cybersecurity strategies and cooperation with others as the essential factors to protect themselves. For example, Singapore has a good cybersecurity relationship with Southeast Asia countries like (Laos, Indonesia, and Malaysia); and two other strong cybersecurity countries such as Japan and Australia. In fact, Singapore signed a Memorandum of Understanding (MOU) about cybersecurity with France, India, the Netherlands, the UK, and the USA by the Cybersecurity Agency of Singapore (CSA). Moreover, India is one of the Asia countries which has cybersecurity cooperation with Malaysia, Singapore, and Vietnam. Especially, there is only Vietnam signed MOU with one of the V4 group – the Czech Republic on cybersecurity cooperation on 14th of April, 2017 in Prague [45]. As a result, some Asian countries are now recognizing the dangerous impacts of cyber threat attacks and concentrating on figuring out the cybersecurity strategies not only themselves but also through cooperation together in order to fight against the global cyber threats. Europe countries focus on political sharing information and cooperation level; however, Asian nations mainly cooperate or share knowledge platform for finance in several organizations such as The Financial Services Information Sharing and Analysis Center (FS-ISAC), Kroll company, High Technology Crime Investigation Association (HTCIA) in Hong Kong, and so on [46], [47], [Table 1.4]. These organizations have strong cooperation between the members in order to handle and respond to cybercrimes, especially in fraud and financial crimes because they were supported by the private sector in finance and intelligence. Significantly, in the EU and the USA nations, they have cooperation in the protection of the secure e-mail system between them. Although in ASIA, there is non-state network in political cooperation, V4 and South East Asia are similar in the intention of responding to cyber attacks in real time on the financial system.

Table 1.4: Several Asian organizations in Finance cooperation

Organization	Functions
FS-ISAC	- Sharing and analyzing for cyber and physical threat intelligence - Sharing incident information between financial services firms worldwide - Warning security to multiple members - Investigating threats and giving recommended solutions - Offering training courses for safeguarding company against security threats
HTCIA	- Providing education and collaborations to prevent and investigate cybercrimes - The global association for cybercrimes detection - Sharing information, skills, and techniques within an association - Investigating professionals within private and public businesses - Identifying and using best practices to gather digital evidence - Building laws and protection for infrastructure and economy - Evaluating the truth by using effective techniques within the digital information
Kroll company	- Monitoring, detecting, and responding to threats virtually anywhere - Securing assets and people - Offering some services such as security disaster planning, policy and procedure development, staffing studies, and so on

1.2. CYBER-CRIME

There are several terms related to cyber-crime like computer crime, information technology crime, or high-tech crime [48] [49]. In the past, cyber-crime was considered with two major categories such as the computer as a target of the attack and the computer as a means of attack. Firstly, the computer as a target of the attack - the attackers use some special tools in order to get unauthorized access and illegally manipulate the confidentiality, integrity, and availability of data. Secondly, traditional offenses with the assistance of computers, computer networks, and communication technology. For example, the blackmailers use the computer to spread out a thousand blackmails or spam messages to the victim computers. Moreover, cybercrime offenses have also ranged from economic offenses like fraud, theft, terrorism, extortion, etc. On the other hand, cybercrime includes some non-money offense activities as programming viruses, spam, and spyware on the computer network or posting some confidential business information on the Internet [50]. The current cybercrimes are no such different from traditional criminals because their purposes want to make money as fast as possible. However, the current computer crimes are more sophisticated than the old ones with many forms on the Internet like child pornography, copyright or trademark infringement, money laundering, cyberbullying, online gambling, and so on [48]. As a result, cybercrime is currently separated into two main categories: machine-made attack and man-made attack [APPENDIX 3]. Machine-made attack defines some cyber-attacks by using computer network environment as a tool to exploit illegal sensitive data and sabotage them, especially in financial damage. In contrast, a man-made attack is considered as a cyber-terrorist attack by an individual or group of people with the purpose of politics and military. Two categories are listed below:

1.2.1. Machine-made attack

Hacking / Unauthorized access to a computer system or networks
According to (N. LEENA), cybercrime is sophisticated, especially hacking the system. Hacking refers to the illegal access activities through a computer or network without authorization to take the privileged access right for all data or systems [50]. Moreover, a computer crime is normally referred to as hacking activities by applying some tools via the Internet to log into the system or break the system just for the challenge, reputation, or profits. Hackers use powerful tools such as keylogger, Trojan, spyware, etc. to poison the victim's computer and take the user's account. Then they try to approach the privilege right of network or computer administrator. Attackers can access illegally to all data in the system, destroy the whole system without notification.

Data diddling
Data diddling is an unauthorized altering of data at various points along the chain of the information entered into the computer system. It can manipulate the output of data and it is not easy to identify. In another hand, these shifts can happen before or during data input or before output. This type of crime refers to bank records, payrolls, inventory data, credit records, school transcripts, telephone switch configurations, and virtually all other apps of data processing [52]. In addition, this type of change can be influenced by someone belonging to the process of creating, recording, encoding, converting, and transferring data that come in a computer [53].

However, we can use some cyber forensic tools which we can trace out when the data was changed and changed back to the original form.

Web jacking

Web jacking is a technique that hackers create a fake website to deceive the victims. When the victims click on the link to the website, it will appear a message that the website has moved to another, and need to click another link. If the users click on that link, it will redirect to a fake page. In another hand, the web jacking process happens when the users connect to a trustworthy domain name that is tricked by Web-Jackers. It is usually done for money, political objectives, and taken the user's credentials. E.g. in 2000, a Web-Jacker stole the web.net domain name which was registered by a small Internet service provider to 3500 non-profit companies [54]. Moreover, based on the report of cybercrime statistics and trends in 2017, there were more than 600.000 Facebook accounts are compromised every day [55]. Each in 10 people who use social networks said that they are the target of a scam or fake link on social network platforms. Therefore, this kind of attack is a common type of cybercrime to get the sensitive data of users.

Salami attacks

Salami attack is also known as another name - salami slicing. This type of crime normally occurs for committing financial crimes. An essential feature of this type is that when small attacks add up to one major attack and it is very hard to detect. This type of attack usually refers to the bank sector and the consequence of the first attack is negligible; however, it happens continuously many times, as a result, the impact is unpredictable [56]. Moreover, a salami attack can be an insider attack (the person who knows the system) or an outsider (the others from outside network system)[57]. The attackers targeted bank holders or individuals who often use online banking or internet banking to make a transaction. Hackers use some special tools to achieve the victim's account information during a money transaction. In fact, in 2008, a man (Michael Largent, 22 years old, Plumas Lake, California) was arrested for collecting money from 58000 accounts through verification deposits from online brokerage firms a few cents at a time [58].

Child pornography

Child pornography includes the creation, distribution, or accessing of sexual materials (photographs, videos, and audio recordings) which involve a prepubescent person. There are some levels of pornography image such as indicative, nudist, erotica, posting, erotic posing, explicit erotic posing, explicit sexual activity, assault, gross assault, and sadistic/ bestiality [59], [60]. Child pornography has many negative effects on child victims such as physical injury and pain, headaches, sleeplessness, nightmare, depression, the feeling of shame or anger, sexually transmitted diseases, and the like. Moreover, pornography not only affects the individuals but also family, marriage, and community. For example, men who usually watch pornography, have a higher tolerance for abnormal sexual behaviors, sexual aggression, and even rape [61]. Furthermore, men begin to see women and children as sexual objects for their pleasure, as a result, it easily leads to sexual harassment or sexual assault, and sexual crime. In addition, pornography can be addictive and cause negative consequences such as marital dissatisfaction, losing emotion with a spouse, or even divorce. Marriage men or women who are involved in pornography feel less emotional or satisfied with their real sexual intercourse or sexual relationship. Last but not least, black websites or pornography websites usually contain potential risks and vulnerabilities. Hackers take advantage of embedding some malware, bad codes inside the pornography images, videos, and links to capture confidential information of the users. They targeted the curiosity of users, adolescents, adults, or viewers from pornography pictures or videos in order to penetrate the computer system via the Internet.

Spoofing and Phishing

Spoofing – means pretending another individual to make a telephone call or sending out emails that present to be someone they are not, i.e. phony name or company [62]. Phishing – creating websites that look like a bank or other business company. Then the phony website requires you for sensitive data (password, credit card, etc.) to gain access to these important personal data. In fact, since 2003, the report said that most of the big banks in the USA, UK, and Australia have been attacked with phishing attacks [63].

1.2.2. Man-made attack

Money laundering

Money laundering is the process of money transfer from crime profits or business crimes. This term is defined as the funneling of cash or other properties from illegal activities through legitimate financial institutions to cover the source of funds [64]. In other words, it involves the activities and financial transactions that are attempted to hide the original source of income [65]. Regarding [P. Reuter and E. M. Truman], money laundering has three vital elements as placement, layering, and integration. The placement state means the physical currency's movement from illegal activities to a place that is authorized and easy to the criminal. The layering state is related to financial transactions as wire transfers to hide the proceeds. In the last stage, illegal proceeds are converted into lawful business earnings through normal financial operations. For example, a businessman – Robert Maxwell used the New York Daily News as his money laundering device with approximately 240 million dollars during nine months working there [66].

Fraud and financial crimes

Fraud is a term which refers to give distortion of thing and money laundering [64]. The Internet brings a good environment in the global marketplace for business and customers. Besides, it also has advantages in anonymity and speed, as a result, attackers may use these factors to make fraudulent activities online. In fact, in 2010, the Internet Crime Complaint Center (IC3) indicated that there are top ten cybercrimes such as non-delivery payment merchandise, FPI- related scams, identity theft, computer crimes, miscellaneous fraud, advance fee fraud, spam, auction fraud, credit card fraud, overpayment fraud [50]. Identify theft is related to all kinds of crime in which someone misuses another personal data like bank account number, credit card number, and telephone number in fraud ways for the economic purpose [67]. In addition, credit card fraud is a kind of crime that someone picks up the other's credit card or pretends gained account information from illegal intrusion and uses it for benefit in e-commerce [68]. For example, in 2016, regarding on IC3 report, there were nearly 15,900 victims all over the world from credit card fraud and it damaged approximately 48,190 million dollars [69].

Online gambling

With the booming of the Internet, online gambling becomes a way to get a huge amount of money from the business, it can attract a large number of users over the world. Although internet gambling is legal in 85 countries in the world, it is an illegal way to conduct financial transactions online in the USA [70]. Because online gambling involves a huge volume of transactions, cash flow which is easy for money laundering [71]. Moreover, online gambling has many negative effects on individuals such as leading people to lose track of time, decreasing the perception of the value of cash, loss of control, legal problems, financial ruin, loss of career and family, and so on [72]. In fact, according to the report of the American Gaming Association (AGA), there are nearly 3000 Internet gambling sites that have turnover approximately 30 billion USD in annual revenue. In another way, on online gambling websites, there are a lot of advertisements that have potential security concerns, as a result, it is a hot pot for hackers to exploit to capture user's data information during the transaction execution when they access the gambling websites.

Data alteration or data theft

A popular type of computer crime that has the main purpose makes illegal changes or steals data. It is related to the integrity of the data. Attackers use special techniques to exploit and penetrate the victim's computer system. Typically, this process occurs during three stages such as acquisition, using, and discovering [73]. In the first stage, hackers want to gain the information from the victim's computer via computer hacking, capture packets during a transaction online's process on the Internet. In the second stage, after gaining useful target information, hackers use them for financial profits. For example, with all information of the target's bank account number with the username and password, they can use this information to illegally purchase online. In fact, according to the US Postal service, there were around ten million identity theft incidents in 2004, which damaged around 5 billion dollars for consumers [74]. In another way, hackers can alter the content of data like user's password, school transcripts, bank records, and the like in order to block the user to use them anymore. In the final stage, even though there is a lot of misuse of credit cards are found quickly, it may take a long time (approximately 6 months to some years) to discover data theft. In addition, the longer the theft is discovered, the greater the damage to the victim who may not involve in law policies in using their sensitive data on the Internet.

Email bombing

It is a kind of denial of service attack - an email bomb that includes a lot of emails to an address in order to overflow the mailbox of the receiver or overwhelm the server. For example, in 1996, a report described a 21-year-old university student at Monmouth, the US who used a mail bomb to jam computer mailboxes of students, staff, and administrators to send and receive messages [75]. However, it is designed simply and easily to detect by spam filters. Moreover, in 1998, in a war against the Sri Lankan government, the rebel Tamil Tigers use an email bombing attack on government servers. It attacked 800 emails a day to Sri Lankan embassies in two weeks to interrupt the communication. Recently, there were over 100 email addresses in the US government were attacked with an email bombing attack in 2016 [76]. Email bombing has a variant - ZIP bomb. Nowadays, some commercial email servers like Gmail, Zimbra, and so on have integrated antivirus software to check and filter malicious file types, Trojan, the virus that compressed into archives as ZIP or RAR. As a result of that, black hackers use another method that they can create an email bomb with the content consisting of enormous text file for an instant, only one letter "a" repeated millions of times and zip it into a small archive; however, when its unpack, it can cause result in denial of service by using a high amount of processing power, RAM (especially for an early version of email servers). With the new technology, modern email server computers have become smarter to recognize such attacks without interruption of service.

Cyberbullying

It is related to changing the images, sending threatening messages, and terrorizing someone. This term refers to a deliberate [77], repetitive and permanent behavior pattern against a defenseless victim by a group or individual via text messages, pictures/clips, email, chat, and websites [78].

Nowadays, with the booming of ICT and social networks, the young generation can send some distressing messages via smart devices like smartphones or tablets in order to humiliate others. Moreover, some teenagers can use mobile phones to take photos, make videos in the bedroom, bathroom, or other places where privacy is expected [79]. Furthermore, in recent days, a serious case is that some teen couples when they said goodbye to each other, used their porn photos or videos and posted them on web pages, social networks for the world to see, tag, and discuss [80]. The negative effects of this crime are both physical and mental for the victims. The sufferers maybe lose their confidence; feel embarrassed or afraid when they face their friends, family, and society.

Steganography

Steganography is the art and science of invisible communication [81]. The word steganography is original in Greek as "covered writing". Steganography and encryption are both used to ensure data confidentiality. However, the main difference between them is that encryption anyone of both parties can see cipher-text when they communicate in secret. Steganography hides the secret text and no one can see that even both parties communicate in secret [82]. With this type of cyber-crime, it can offer more chances for attack than the marking technique itself. For example, Digimarc – a best-known for its digital watermarking technologies company was attacked by using a weakness in the implementation [82]. During the user's registration process with the marking service, a debugger maybe break into the software with checks these passwords and disable the checking. Then the hackers can change the user's ID and this will change the mark of existed marked images in the system. It may allow bypassing of the checks to see if a mark existed; therefore, it enables marks to be overwritten.

Computer vandalism

Computer vandalism is a type of malicious behavior that is related to computer sabotage and data in many different ways. They differ from viruses because they can attach themselves to existing programs. Typical damage of this type is erasing hard drive data or extracting login credentials [83]. There are four major types of vandals as talented students, amateur youths with the assistance of the Internet, expert developers, and researchers. Firstly, in some cases, some skilled students who have a strong passion for a computer programming language, want to figure out their abilities or their skills and show off themselves by creating some malware codes and send them to the network for victims. In fact, on 27th November 2006, the website of the Ministry of Education was attacked by a 17-year-old student in Vinh Long, Vietnam [15]. He exploited this website's security holes and changed the Ministry of Education leader's profile picture as a means to prove his skills. Secondly, an amateur young generation is not quite good at coding; however, they prove their self-confidence in making viruses or malware. Because of the Internet environment and self-study websites, these individuals can create and distribute their own viruses via the Internet. Thirdly, with the expert developers, they are mature and they can make many complicated programs that use the latest methods to penetrate the data system or take advantage of security vulnerabilities. Latterly, with the purpose of research, the researcher's group as an ethical hacking team invents new methods to infect computers system to find the potential vulnerabilities in order to create antivirus software. However, these methods may be used by bad intentional people or criminals.

However, there are some cyber-crimes in both machine-made attack and man-made attack such as hacking; spoofing and phishing; email bombing, and unauthorized access to a computer system or networks.

Cyberterrorism

Cyber-terrorism is a combination of cyberspace and terrorism. It refers to the attacks against computers, networks, information and the consequence of the attacks terrify the government, political and social objectives [84]. In another way, cyber-terrorism considered as politically motivated computer attacks toward other computer systems that lead to threatening victims of attacks [85]. There are several kinds of cyber-terrorism such as attack can lead to death or bodily injury; some can damage the critical infrastructure or economic loss. Cyber-terrorists belong to a funding organized group for their activities, so they can hire a lot of hackers to act on their behalf [85].

Cyber-extortion

Cyber-extortion expresses the criminal money requirement activities or exchanges some valuable things in order not to spread out the threats to computer users [86]. Nowadays, ICT is becoming more central and essential to everyone, companies, therefore, cyber-extortion are more sophisticated, well organized, and dangerous for not only individuals but also for companies. For instance, on March 1, 2004, four people were arrested for trying to extort two billion yen (approximately 18 million USD) by threatening to release nearly 4.6 million customers' sensitive data from Japan's leading broadband Internet service provider (ISP). In another case, on November 23, 2003, Mickey Richardson – the boss of an online gambling website called "Betcris.com" in Costa Rica, received an email with the message "Your site is under attack and you need to send 40 thousand USD and your site will be ok for next 12 months unless your site will be under attack continuously during next 20 weeks until you close the doors [86]". According to the Internet Crime Report (ICR3), there was a slight decrease of the victims from 17,804 to 17,146 but the damage increased from 14.7 million around 15.8 million USD between 2015 and 2016, respectively by cyber extortion [69],[87].

The connection between cybercrime and cyber-warfare

Cybercrime and cyber warfare have similar purposes such as political interest, finance, military, or other aspects as a religious or social ideology [88]. Therefore, the boundary between cybercrime and cyberwarfare is slightly blurred. In addition, hackers and terrorists have similar interests as well. They are easy to get motivation from profits which comes with organized crime and can be sponsored as terroristic groups or countries. In fact, North Korea is an illustration country in the development of cyber warfare from cybercrime. For example, according to the information security firm, North Korea stated a bigger threat of large-scale cyber-attacks than Russia in 2016 [89]. Moreover, in recent years, North Korea has been linked with a series of online attacks on financial networks in the USA, South Korea, and other countries [90]. Especially, some cybersecurity researchers also said that they have found the global "Wanna-cry ransomware" attack in 2017 which is related to North Korea. It infected and damaged more than 300,000 computers in 150 countries over the world. In short, all mentions above showed that the relationship between cybercrime and cyber warfare; hackers and terrorists as well are really tight.

1.3. CYBER-WAR

Cyber-war or cyber warfare is a combination of computer network attack and defense by using special technical operations [91]. In another way, cyberwar is considered as an action that uses ICTs within an offensive or defensive military strategy endorsed by a state in order to immediately disrupt or control the enemy resources [92], [93]. In addition, "cyber warfare is also the art and science of fight without fighting; of defeating an opponent without spilling their blood [94]." Furthermore, cyber warfare at the government level mainly focuses on political, cultural, and military situations in another country as a target or for specific offensive or defensive operations in cyberspace [95]. Although there are many definitions of cyber warfare, in my opinion, it mainly focuses on achieving military objectives during the war between two countries or with the other countries.

Espionage

Cyber espionage considers as an act to steal secret information or private data from individuals, organizations, and government for personal, economic, military, and political purposes by using malicious software such as Trojan horse and spyware. A good example of cyber espionage is the Stuxnet virus in 2010. It was designed to control and monitor the physical hardware of Iranian nuclear facilities [96]. This kind of virus was extremely sophisticated because it could damage the physical hardware. Moreover, there are three other major espionage tools that seem similar to Stuxnet (Gauss in 2012, Flame and DuQu which steals passwords; monitors the computer's keyboard and network traffic; and collects data, respectively [96][97]. Due to the complication and similarities of these viruses, the researchers believed that they were created by the United States or Israel, even though neither of them claimed responsibility for that. Nowadays, cyber espionage plays an essential role in cyber-attacks, there are many countries take advantage of this method as a powerful tool for cyber warfare such as United States, Russia, and China. In fact, Russia used Moonlight Maze virus to steal private information from Department of Defense, Department of Energy, National Aeronautics and Space Administration (NASA) and military contractors of United States in 1999 [98]. Moreover, Russia used the DDoS attack on Estonia to stop services of important websites and disrupt communication across the country in 2007 [99]. Rather, cyber espionage can be used to sneak the information in economic and financial as well. For example, the United States' economy can lose from 25 billion dollars to 100 billion dollars annually from Internet hacking because of the loss of financial data by Chinese hackers[100]. Therefore, the effect of cyber espionage is extremely high and dangerous. This attack can limit or block the victim nation's ability to defend, it can lead to the loss of property, communication system, critical infrastructure, and citizen life.

Sabotage

Sabotage is considered as a malicious act that can interrupt the normal processes and functions of the system or damage the equipment or data in the system. In fact, in November 2007, Seagate Maxtor Basics Personal hard drives were exploited with a Trojan horse virus. This kind of Trojan was created to copy data on the computer and send it automatically to Beijing websites [101]. Moreover, sabotage is an intentional effort to destroy or reduce the strength of the economy or military system [102]. For example, on 6 September 2007, it called Operation "Orchard" when Israel attackers used electronic warfare in taking out and disabled the radar of Syrian's air defense system in order to use Israeli squadron of F-15I and F-16I warplanes to enter Syrian airspace. Even though this cyber-attack didn't destroy physical anything, it was also considered as a successful attack on the military operation of the enemy. Moreover, in a report of Symantec, there was particular cyber sabotage that happened on 23rd December 2015 in western Ukraine. The hackers used the malware namely BlackEnergy Trojan (Backdoor. Lancafdo) and Trojan. Disakil to gather information and break the critical electricity systems [103].

Denial of service attack

Denial of service attack (DoS) is an attack that interrupts the victim's service. While the attack happens, customers cannot be able to use any services from the victim's website. DoS attacks are based on the weaknesses in the IP protocol stack to disrupt Internet services [104]. There are some kinds of DoS attacks such as against users, hosts, and networks. Normally, a DoS attack is related to an individual attacker who can take advantage of vulnerabilities of the victim's computer, break into target servers and then bring the system down [105]. Moreover, a normal computer individual can be a DoS attacker as well with tools from the Internet easily e.g. Trinoo [106]. However, it is difficult to overwhelm the target's resource with a single computer; therefore, the attackers need to use a large number of distributed attacking hosts on the Internet – namely distributed denial of service (DDoS) attacks. These host like a zombie (an assistant program which connects directly to master hosts) will wait for the command from the attacker and amplify the signal to attack the victim. They can generate hundreds of megabits per second signal floods in order to send many packets to the victim's server at the same time. It leads to the victim's system out of service [Figure 1.3]. Compared with traditional DoS attacks, this attack is more powerful and complex. The DDoS attack has 2 stages: creating a zombie and attacking victims. Firstly, attackers need to infect a large number of hosts on the Internet by exploiting the vulnerabilities of the victim's system and sending some malicious code via malware, Trojan, cracking apps, etc. For example, hackers can create a small free game and put the Trojan inside. After that, they upload that software on the Internet and wait for the user who downloads and installs them into their computers. Therefore, anyone who downloads and runs that software becomes an unwilling zombie and waits for the commands from master hosts. Then, attackers use few commands to communicate with zombies via (DNS, ICMP, HTTP, and IRC), wake up them, and launch massive attacks against to victim. In fact, regarding the Kaspersky lab report, there was a heavy DDoS attack against some of the largest Russian bank websites from 8 to 12 November 2016 by many bots from 30 different countries which were from the United States, India, Taiwan, and Israel [107]. DDoS attacks are quite more popular and they become a major threat to all public services in the world. Because when the attack occurs, it may block a huge amount of hosts by sending flood data packets and make the system down.

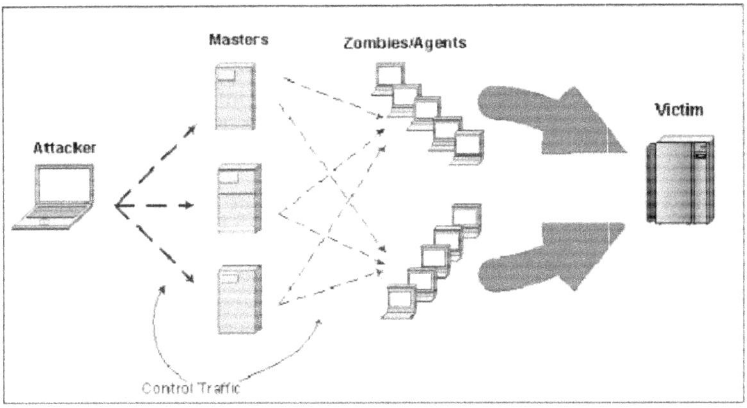

Figure 1.3: Description of DDoS attack [104]

Regarding the definition of cybercrimes above, there are some kinds of cybercrimes that have a significantly strong effect on national security at the governmental level (military, bank, and critical infrastructure). However, some of them dramatically influence on citizen level [Table 1.5].

Table 1.5: The effect of cybercrime on a governmental level and citizen level

Type of Cybercrime	Military		Bank		Critical Infrastructure		Citizen	
	Weak	Strong	Weak	Strong	Weak	Strong	Weak	Strong
Transnational crimes		x	x		x			x
Child pornography	x		x		x			x
Data diddling		x		x	x			x
Salami attack	x			x	x		x	
Web jacking		x		x		x		x
Hacking/Unauthorized access to a computer system or networks		x		x		x		x
Spoofing and phishing	x			x	x			x
Money laundering	x			x	x			x
Data alteration or theft		x		x	x			x
Email bombing	x		x		x			x
Cyberbullying	x		x		x			x
Online gambling	x			x	x			x
Steganography	x			x	x			x
Computer vandalism		x		x		x		x
Fraud and financial crimes	x			x	x			x
Cyberwar		x		x		x		x
Espionage		x		x		x		x
Denial of service	x			x		x	x	
Sabotage	x			x	x			x

Fake news or disinformation

Currently, there is a new type of cyberwar by social engineering attack (Fake news or disinformation). Fake news and disinformation are closely similar; however, all fake news is disinformation but several disinformations is fake news, and the entire is a misrepresentative reality [108]. Moreover, fake news or disinformation includes false or misleading information in order to falsify or harm the public with financial or political motivations [109], [110]. This type of cyberwar has several complex impacts on the digital information system and democracy systems such as reducing the levels of trust in institutions and experts, and increasing the doubt in people's thinking. Indeed, in 2016, one day before the US presidential election, there were some messages on social media (Facebook, Twitter) said that Hillary Clinton had died and this message made the voters believed that the date of the election had changed [111]. In order to handle this type of cyberwar, the EU Commission offers an action plan to safeguard the European values and democratic systems, following by enhancing transparency regarding the way information created or supported; a variety of information; reliability of information; and comprehensive solutions with broad stakeholder involvement [109].

1.4. CYBER-CRIME VS CYBER-WARFARE

Cybercrime is a crime that involves computer technology to access sensitive data, malicious purposes, and illegal activities. There are two types of cybercrimes: computer as a target of the attack and computer as a means to attack. In contrast, cyberwarfare is an act that involves offensive and defensive activities [APPENDIX 4].

Method to attack

Cybercrime uses computers, malicious codes to make viruses, Trojan, malware, and so on to do the attack the target. Meanwhile, cyber warfare combines weapons with high-tech tools to attack.

The consequence of cybercrime and cyberwarfare

Cybercrime may cause several dangerous impacts to the target, following by:

- Strong influences on e-commerce such as integrity, authentication, availability, and authorization privacy during a business transaction.
- Cause of financial damage and monetary losses.
- Influencing the online and offline world.
- Negative effects on the business of both small and big companies
- Major effects on piracy of the entertainment, music, and software industries
- Spending a lot of money on building a security system

While cybercrime's effects mainly focus on the economy and finance; cyber warfare also causes some hazardous consequences in politics, critical infrastructure, and security for both nation and citizens, such as:

- Effecting on politics, national's stability, and citizen's life.
- Damaging critical infrastructure (electricity power grid, water supply, transportation, control system, and so on) of a country.
- Major effects on political and military communications remotely from anywhere in the world
- Corrupting weapons of the enemy
- Effecting on health, security, or the economy, functions of government, and social wellbeing of the population

1.5. GLOBAL TRENDS IN CYBER STRATEGIES, CYBER SECURITY IN COOPERATION

Global security has five major pillars such as legality, technicality, organization, capacity building, and cooperation [112]. Firstly, legality is related to the existence of legal institutions and frameworks which handle cybercrime and cybersecurity. Secondly, technicality depends on the number of technical institutions and frameworks in order to solve cybersecurity issues. Thirdly, an organization is evaluated on the total of policy coordination institutions and strategies for cybersecurity development at the national level in every country. Fourthly, capacity building and cooperation rely upon the research and development organization; education and training programs; certified professionals and public sector agencies promoting capacity; and the existence of partnerships, cooperative frameworks, and information sharing networks, respectively. However, the cybersecurity trends and strategy of each continent is quite different. Last but not least, in European Union, there is a new regulation – namely General Data Protection Regulation (GDPR) which was approved by the EU Parliament on 14 April 2016 and was implemented on 25 May 2018 [113]. This new regulation allows users to control their personal data during processing via the internet environment inside the European Union. Moreover, this regulation also requires any companies, organizations, or websites to make clear the purpose of data collection, show the lawful basis and indicate how long data is kept or shared with other third parties inside or outside of the EU [114]. Although GDPR is applying in Europe, there is one Asian nation that also implements this new regulation on data protection (Japan). In fact, on July 17th, 2018, Japan was successfully deal with the European Union in exchanging, collaborating, and accepting each other's data protection system [115].

ASIA

Almost the cybersecurity cooperation between Asian countries is major in economic, military, and diplomacy. Moreover, the cyber security of the Association of Southeast Asian Nations (ASEAN) member countries lacks cyber capacity building. In fact, regarding the FireEye, Advanced Threat report for the Asia Pacific, cybersecurity capacity building on Asia has not developed well yet [116]. Furthermore, the obstacles to the success of cybersecurity practice and policy implementation depend on ASEAN decision-making in a long process. Because ASEAN countries are based on individual national governments and each national government is different in its ability and in doing [117]. Besides, some Asian countries have started the cybersecurity capability like Japan, China, South Korea, and North Korea but the others are still behind.

EUROPE

Currently, almost all European countries use the common cybersecurity capacity model - the cybersecurity capability maturity model (CMM) [118] [119]. This model includes five dimensions of cybersecurity capacity building:

- Creating cyber-policy and strategy;
- Boosting responsible cyberculture within society;
- Designing cyber-skills into workforce and leadership;
- Building effective legal and regulatory frameworks;
- Managing risks via organization, standards, and technology;

However, each country in the EU has applied a cybersecurity maturity strategy in various ways. Some of them have differences in the concentration and aims such as some countries target on developing the economy and business when others focus on the encountering against cybercrime and the accelerating cyber-defense plans.

In another hand, EU members also focus on building the cybersecurity frameworks [120] as the key factor in order to face cybersecurity challenges.

- Creating legal foundations for cybersecurity
- Building operational capabilities in improving cyber resilience
- Cooperating public-private partnerships
- Making sector-specific cybersecurity plans
- Increasing education and awareness

1.6. WHY DO NEED AN URGENT PROPOSAL FOR VIETNAM CYBERSECURITY STRATEGIES?

In Vietnam, there is no precise legal foundation or cybersecurity strategy, public-private sectors, and sector-specific cybersecurity plans; however, it has only one operational entity - VNCERT (Vietnam Computer Emergency Response Team) which was established in 2005 [121]. Moreover, there is a draft of information security law including some regulations, the requirement for information security and the responsibilities of VNCERT. It was proposed in front of the Vietnamese parliament in 2015 and applied in 2016 [122]. Even though Vietnam's Information communication technology (ICT) human resources are rich, their information technology (IT) professional skill is not enough to well compete with the other countries in the same region and in the world [123]. Besides, the online legislative framework; for example, legal laws or regulations for e-business, e-government, e-marketing, and the like didn't get completely [124]. In addition, the current internet service providers (ISPs) also skip the security standards of their networks; hence, the computer security and information assurance issues are a major challenge for Vietnam ICT development not only for officials and providers but also for users [125]. In another way, ICT training projects for staff, workers, and citizens are not paid attention to; as a result, the qualification and capacities of IT staff are at a low level [126]. Furthermore, ICT is expanding with incredible speed in Viet Nam, especially Internet users; however, the threat of cyber-attacks in critical infrastructure also increases quickly. It threatens not only national critical infrastructure but also national security and citizen's life. The attackers mainly use the Internet as a powerful environment to hijack some parts of critical infrastructure as the government agencies, industry, and transportation. These damages influenced Vietnamese critical infrastructure, especially in air transportation. In short, the Vietnamese government needs to invest more budgets in some IT training projects not only for organizations but also for individuals in order to upgrade IT skill. As a result, the Vietnamese government requires urgent cybersecurity strategies for defending against global cyber threats.

1.7. SUMMARY

This chapter has presented basic concepts of two major types of cybercrime – (man-made and machine-made attacks) in the Internet environment, the differences and impact between cybercrime and cyberwar on national security. Besides, it can be seen the security and economic cooperation among the Asian nations. Some of the Asian countries have bilateral cooperation with the USA like US-Japan, US-South Korea, US-Australia, US-Philippines, US-Thailand; some with China such as Russia-China, India-China, USA-China, North Korea-China, Japan-China. Likewise, there are also trilateral cooperation nations such as Japan-South Korea-USA, Russia-China-USA, Japan-India-USA, and Japan-China-USA. Generally, almost the cooperation between South East Asian nations is in the trade, energy, peace, and friendship cooperation but there are few nations that cooperate in cybersecurity. Among them, Singapore and Japan are the leaders and the second in the same region in cybersecurity cooperation, respectively. Significantly, Vietnam is the only nation that cooperates in cyber security with the Czech Republic (one of the V4 countries). However, Asian cybersecurity cooperation mainly focuses on sharing the information and knowledge to protect the economy through several private sectors in finance and intelligence like FS-ISAC, Kroll Company, and HTCIA. Unlike the European Union and because of a non-state network in political cooperation, the author strongly believes that Singapore's cybersecurity strategy can be adapted for Asian countries to follow, especially Vietnam. In another hand, I strongly believe that Cybersecurity in the East Asian and South East Asian countries aim to create a more secure society and supports economic development and Singapore's cybersecurity strategy can be adapted for Vietnam's legal framework. In the next chapter, there is a description of cybersecurity, policies, and strategies of Visegrád countries and how cybersecurity works.

Chapter Two:

CYBERSECURITY, POLICIES, STRATEGIES, COOPERATION IN VISEGRÁD COUNTRIES

In Europe, each country has different strategies to ensure its national security, especially in cybersecurity. As each country has its own contexts, strengths and technology development, and policies, it may be difficult to cooperate and operate the same strategy. This chapter, therefore, reports on how to collect data, analyze and compare security information from Visegrád countries, the EU, NATO, and other organizations by taking the consultation from cybersecurity experts and regarding the V4's official legal framework and national sources to identify the differences and similarities among Visegrád countries' strategies. Furthermore, regarding data collection and data analysis from these sources, it was expected to find the answers to how to ensure Visegrád's power in the same region. Likewise, it was also expected to explore European countries' legal frameworks or organizations to promote cyber defense policies and ensure the security of cyberspace for the Member States.

2.1. GENERAL POLICIES, STRATEGIES, COOPERATION OF VISEGRÁD COUNTRIES

History of Visegrád countries

In the 14th century, there was a meeting of three kingdoms for an agreement in alliance treaty – John of Luxembourg, Charles Robert, and Kazimir III from Czech, Hungary, and Poland in Visegrád on Danube River (Hungary), respectively [127],[128],[129]. The goal of the summit meeting was related to the cooperation of trade, taxes, and trade routes. Moreover, this cooperation was a mini model for the future of the European Union. After nearly one hundred years of cooperation, lots of events between Visegrád countries occurred. Significantly, after Warsaw Pact, three representatives of three countries first met in Bratislava on April 9, 1990, and then they signed the Declaration on cooperation between them with Slovakia on 15 February 1991 in order to join for European integration – considered as V4 group. This foundation – V4 group contributed to the establishment of the Central European Free Trade Association (CEFTA) on 21, December 1992. This agreement was considered as a successful project in the economic transition between Central and South-West Europe; however, almost all members left after joining the EU. After the long inactivity period of V4, in May 1999, the content of the Visegrád Cooperation document was approved during the V4 Prime Ministers summit at Bratislava. Regarding the Visegrád cooperation document, every country in V4 went around in a circle of Presidency every year. Moreover, there was an official and unofficial summit of the V4 Prime Minister; governmental, diplomatic, and expert meetings in the Presidency country. After the success of the CEFTA's establishment in 2000, V4 reached the breakthrough step which was the creation of the Visegrád International Fund – VIF in Bratislava [127]. This financial fund is essential in supporting many security projects; however, these projects are small and their importance in the area of civil security also has a minor impact towards the international scope [130].

Reasons for Visegrád cooperation

Visegrád countries have several similar factors such as historical development, culture, economics, society, geography, and security problems [127], [131], [130] of the region in order to have natural cooperation for potential enhancement for the Central Europe area. Each country of V4 perceived the importance of cybersecurity issues; cyber threats and urgent needs to protect their citizens' security. Moreover, they maintain their democracy, citizen's rights (freedom to talk and access to information), confidentiality information, and privacy [132]. Furthermore, Visegrád countries are the small countries in Eastern Europe with a total surface of around 500.000 km^2 and 60 million citizens, nearly equal to the surface and population in France. On the other hand, the number of V4 armed forces is about 200.000 which is similar to German or United Kingdom [127]. Besides, the most important key factor of cooperation is that entering the European Union in 2004 and other international organizations. In fact, the "back to Europe" is the V4 group's slogan when it was established [130]. In addition, Visegrád countries expected to have an organization which represents intensive relations among these countries, especially in regional, historical and cultural similarities as well as the obstacles from former Socialist countries [131]. Likewise, their geographical features, infrastructure, consumption structure, and the ability of capital attraction are similar [128]; therefore, Visegrád's cooperation created a strong regional organization in Europe. It is important to note that the votes of the V4 together in the Council of the EU are nearly equal to Germany and France together. However, the perception of each country about cooperation is quite different. Indeed, Visegrád group citizens believed that economic cooperation is the major reason. Hungary and the Czech Republic considered the EU entry is the second important goal while Poland and Slovakia stress justice and order maintaining as the second one after economic cooperation [131].

The aims of V4 cooperation

Visegrád cooperation aims to have these purposes following by:

- Restoration of the country's sovereignty
- Democracy and freedom
- Liquidation of totalitarian system's residua
- Building up of parliamentary democracy, a modern market economy, and modern legal state
- Full integration into the European political, economic, security, and legal systems.
- Building the European security architecture depended on effective, functionally supporting, and mutually strengthen cooperation and coordination with European and transatlantic institutions [133].

Visegrád mechanism

Every year the presidency of V4 rotates for each country. Each president composes his/her own program to ensure long-term cooperation for V4. At the end of each presidency, there is one official Prime Ministers summit and there are several informal meetings of Prime Ministers and Foreign Ministers before international events. Besides, there are some meetings between V4's presidency and other ministers in V4 and V4+ format. In addition, the role in internal and inter-state coordination of the national coordinators as well as their communication are also improved. Moreover, there are some meetings between Presidents and the Parliament of Visegrád countries in an annual year. V4 has the mission of keeping contact and cooperating with Permanent representations to the EU and NATO in Brussels as well as some organizations like OSCE, UN, COE, OECD, WTO, and so on. Last but not least, V4 needs to enhance the International Visegrád Fund and its structure [129]. Nevertheless, each V4 cybersecurity strategy has different fragmentation itself [Table 2.1].

Table 2.1: Fragmentation authorities of Visegrád countries.

Country	Authorities
CZECH REPUBLIC	- Cyber Security managed by Ministry of Interior (2010 – 2011) -National CERT (CSIRT.CZ) -CZ.NIC –legal entities operation in (domain name, e-communication market) -CERT (GOVECERT.CZ) - Military CERT/ CIRC administered by the Ministry of defense (armed forces, defense ministry) - 20 private CERTs
SLOVAKIA	- National CERT/CSIRT - CSIRT.SK response in the civil sector - Ministry of finance. CSIRT.SK cooperates with a similar team on the international platform on a regional level with the teams of V4 and Austria. - CSIRT.MIL.SK – for monitoring, evaluation, measure-taking of information security
POLAND	- Strategic/policy level by Ministry of digital affairs with the Ministry of finance, justice, interior - National Center for cybersecurity and national CERT/ CSIRT with sectoral CERTs/CSIRTs (energy, financial, banking, water supply, administration...) -Technical level including SOC (security operations center) - Ministry of finance responsible for cybersecurity issues - Ministry of defense used for national security, military security. - Ministry of interior responsible for critical infrastructure
HUNGARY	- At government level (National cybersecurity council supported by national cybersecurity forum, academic & business sector council, some task-oriented workgroups. - Ministry of interior – for central governmental incident management, Critical Infrastructure. - National CIRT (or GovCERT) - Ministry of defense –for military incident management (MilCERT) - National Directorate general – for disaster management - Hungarian internet service providers – providing civil domain - NIIF institute – NIIF CSIRT workgroup to protect Hungarian mid-and higher education and research sector - GovCERT and MilCERT – tend to keep secret instead of sharing data.

2.2. HOW THE CYBERSECURITY STRATEGY FRAMEWORK IN THE EU COUNTRIES

The Czech Republic is a pioneer in the information society in central European countries; however, the national information infrastructure is confronted with cyber threats, especially (well-known case – Stuxnet) [134]. Moreover, cyberspace has no precise geographic barriers between one country and another one; therefore, every state cannot deal with global cyber threats alone, they need to cooperate with international partners to figure out the solutions. Hence, it is essential for each state in European countries to find out the means of protecting information infrastructure to counter global threats. However, the principal questions for states are hereby: how to find appropriate solutions against cross-border cyber attacks and how to cooperate with the countries at the international level. The main key factor of international cooperation is the trust between member states in central Europe. As a consequence, in 2013, the Central European Security Platform (CECSP) consists of Hungary, Czech Republic, Poland, Slovakia, and MilCERT - military CERT of Austria was founded. This forum based on the trust and sharing information between the states in order to work together in the field of cybersecurity. The main purpose of this cooperation is to build up the cybersecurity level for the countries in the same region and contributing significant impacts to the EU and other international organizations (EU and North Atlantic Treaty Organization - NATO) in the cybersecurity aspect. There are several objectives of CECSP's cooperation such as sharing information and best practices in cybersecurity aspects; creating and deploying secure communication channels; defining and simulating a categorization system for sensitive information; reconciling individual positions for the international forum, and creating practical working groups [134]. Firstly, every member state improves resilience and readiness by sharing information and best practices in the cybersecurity field in order to counter cyber threats. Moreover, each state also exchanges the cyber-attack incidents; malware or virus information; potential attacks; research and development projects including education and training under State's voluntary. Secondly, due to exchanging information between member states, they need to create and deploy secure communication channels for reducing eavesdropping and altering the information during transmission. Thirdly, the States have the responsibility to classify the sensitive information and establish the popular standards of the cyber incident importance for common use. Furthermore, according to promulgation, all partners are required to deliberate their national position before the meetings with the international forums like the EU, NATO, the UN, and Organization for security and cooperation in Europe (OSCE), and The European Union Agency for Network and Information Security (ENISA), as a result, it can assist the bilateral or multilateral relationship and discuss cybersecurity strategies effectively. Last but not least, a minimum of two partners in the Member States can create some important practical or cross-border cooperation groups like techniques, operations, management, and policies to improve critical information infrastructure security (CII), resilience, and cooperation together. In brief, the cybersecurity strategies for Central European countries were established as a common declaration to enhance the popular interests of states and improve the multilateral relationship or cross-border cooperation in order to against the global cyber-threats.

2.3. COOPERATION OF VISEGRÁD COUNTRIES ITSELF, WITH EU AND OTHER INTERNATIONAL ORGANIZATIONS

Visegrád countries cooperation within V4

Visegrád countries have cooperation in some areas such as culture, education, science, infrastructure, environment, and youth exchange [135]. Moreover, this cooperation aims to strengthen the civic dimension, cross the border, and Schengen within the International Visegrád Fund and their structures. Additionally, Visegrád countries also expand the transformation's experiences on preventing terrorism, organized crime, refugees, disaster management, and defense industries. V4 also cooperates in managing the disaster, infrastructure, or the environment. Likewise, they enhance the defense and arms industries development to counter back the terrorists. In fact, the EU battlegroup of the V4 group holds regular exercises under the protection of the NATO Response Force. Among this battlegroup (V4 and Ukraine), Polish defense is the leader and the first exercise was held in Poland in 2013 [136]. They also declared the Action Plan in several areas in defense planning, military education, airspace protection, training and exercises, and so on in order to the joint military body within the EU [137].

Visegrád countries cooperation with EU

Beside Visegrád countries cooperates itself, they mainly active the contributions to the EU in order to develop the Common Foreign Security Policy (CFSP), EU strategy towards the Western Balkans, and participate in the development of the European Security and Defense Policy (ESDP) for enhancing the relationship between EU and NATO. Furthermore, Visegrád plays an important role in collaborating on current concerns of common interest, exchanging experiences in Justice and Home Affairs, Schengen cooperation, as well as protecting and managing the EU external borders and visa policy. Similarly, Visegrád cooperation creates new economic cooperation possibilities and forms within the European Economic area and discusses preparations for using the European Monetary Union (EMU). After participating in EU and

NATO, V4 also supported Western Balkans countries (including Albania, Bosnia, Herzegovina, Bulgaria, Croatia, Kosovo, the Republic of Macedonia, Montenegro, Romania, Serbia, and Slovenia) to improve the Western Balkan and their Euro – Atlantic integration process [138]. Indeed, in 2014, there was several practical supporting from V4 for Western Balkan; for example, law, children rights, public, and administrative reform. Remarkably, regarding the migration crisis, there was a meeting between the ministers of interiors from V4, Slovenia, Serbia, and Macedonia in 2016 to improve the control over the migration flows [138]. Besides the cooperation with the Western Balkans nations, V4 also cooperated with the Benelux group (Belgium, Netherland, and Luxemburg) in eight major areas in 2003 to clarify possible common actions such as [139], [140]

- Schengen issues
- Trademark and design office in Den Haag
- The Parliament
- Euro Control Route
- European structural funds, infrastructure, and spatial planning
- Environment challenges and protection, implementation of NATURE 2000, investment policy
- Investment policy, tourism, and promotion in third nations
- Social and labor policy, market issues, cross border employment, employment possibilities

With this cooperation between V4 and Benelux, it can help in increasing the common foreign, security, and defense policy of the European Union. Furthermore, currently, V4 cooperates with Austria and Germany (V4+) to improve stability, reduce the cyber threats to peace and build the security relationship between Euro – Atlantic, and NATO as well as other partner countries.

Visegrád countries cooperation with the other international organizations

Visegrád expands the cooperation with other partners which have similar interests in central European countries, and with EU and NATO. The main aims of V4 cooperation within NATO and other international organizations are strengthening transatlantic solidarity and cohesion, promoting a common understanding of security between the EU countries and Euro-Atlantic, improving the combating international terrorism, exchanging the information in international organizations (UN, Council of EU, OECD, etc.), and consulting in the OSCE on issues of common concern for V4 countries [141].

2.4. THE METHODOLOGY OF THE DEFENSE SYSTEM

Before joining in NATO structure, V4 defense cooperation mainly focused on political consultations. In 1997, this cooperation changed into military and defense cooperation to enhance more effective consultations on defense and security issues among V4 countries. However, V4 military and defense cooperation decreased and turned back to be concentrated on political consultations [142]. After the economic crisis in Europe in 2009, the defense cooperation was as a means to encourage the partners to improve their lacking defense capabilities. Therefore, there was the first document which was signed during the Hungarian V4 presidency, namely "Long term vision of the Visegrád countries on deepening their defense cooperation" in 2014. Visegrád countries cooperation defense strategy focuses on several main tasks such as joint capabilities development, interoperability of the V4 armed forces (education, training, and exercises), and defense industry (participate procurement and acquisition)[142].

- Draft a long-term vision for V4 defense cooperation strategy that would also organize common capability development efforts
- Strengthen cooperation in the field of training and exercises of the armed forces in the V4 format. They envisioned that joint V4 military exercises are organized on an annual basis, harmonized with NATO, EU, and national exercises, as they will provide an excellent tool to increase the interoperability of V4 armed forces
- Explore the possibility to create a framework for an enhanced defense planning cooperation on the V4 level in order to identify new promising areas of defense cooperation among their countries.
- Expand the cooperation with other regional countries like Ukraine, Austria, or Slovenia, as well as Germany or the Republic of Korea.
- Creating the V4 EU Battlegroup to strengthen the V4 defense cooperation in 2011.

This V4 Battlegroup has eight different modules with approximately 3280 soldiers (1450 by Poland, 670 by Hungary, 600 by the Czech Republic, and 560 by Slovakia) [142]. Moreover, each V4 country has its responsibilities on various tasks; for instance, Poland is a leader with the responsibility for training, planning, preparation, communication in an information system. Meanwhile, the logistic, protection from weapons, and engineering are deal with in the Czech Republic, Slovakia, and Hungary, respectively.

In short, the V4 defense and security cooperation created an Action plan of the V4 defense cooperation containing many important subareas on which group members would focus in the future. Moreover, this cooperation is one of the essential topics of the V4 group including the defense planning cooperation, joint training exercises, military education cooperation, as well as V4 EU battlegroup in order to enhance the cooperation with NATO and EU and the contribution of V4 in EU common security and defense policy (CSDP). The V4 military defense cooperation is crucial because it enables the Visegrád group to be recognized more seriously in the international arena.

A new dimension in cyberspace (e-commerce and e-government), cyber defense, cybersecurity strategies in Visegrád countries
With the boosting of ICT, V4 members face many challenges from cyber threats. Hence, Austria and Visegrád countries (V4) began to cooperate in 2013 with the creation of the Central European Cyber Security Platform (hereinafter: CECSP). The cooperation's purpose of five states is to enable the information, best practices, lesson learned, and know-how sharing about cyber threats and potential solutions for cyber-attacks [143]. Moreover, this platform will provide the capacity and capability building in improving the V4 position in the international environment.

2.5. COMPARISON OF STRATEGIES OF VISEGRÁD COUNTRIES AT GOVERNMENT OR TECHNICAL LEVEL

Visegrád countries have a similar history, geography, and culture. Therefore, they want to cooperate to enhance their sovereignty. In general, Visegrád countries have their own cybersecurity strategy with several similarities and dissimilarities [Figure 2.1].

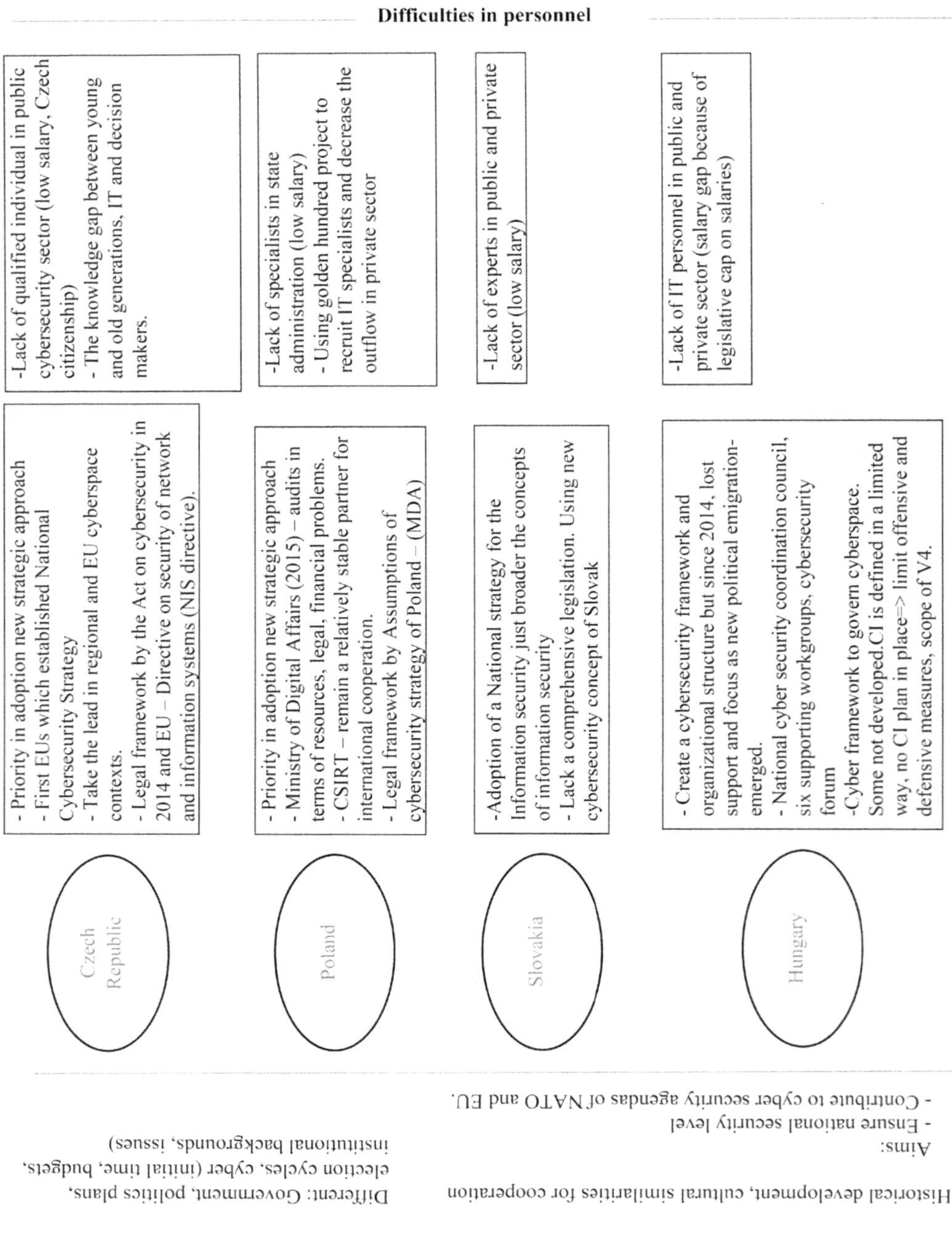

Figure 2.1: V4 cybersecurity strategy

Table 2.2: The legal framework of Visegrád countries [144] [145] [146] [147].

	Slovakia		Czech Republic		Hungary		Poland	
Legal foundations								
	First	Current	First	Current	First	Current	First	Current
National cybersecurity strategy	-National cybersecurity strategy 2008-2013	National cybersecurity strategy 2021-2025	The cybersecurity strategy 2011-2015	The cybersecurity strategy 2021-2025	National cybersecurity strategy 2013	National cybersecurity strategy 2018-2023	Cyberspace protection policy 2013	Cyberspace protection policy 2017
National cybersecurity strategy first applied	2008		2011		2013		2013	
Critical infrastructure protection strategy or plan	-The act of 8 Feb 2011 on the CI – covers the regulation and practices surrounding Slovakia's CI		- The act on Cybersecurity on 1 Jan 2015- provisions for the development of CI plan - Regulation No.317/2014 Coll. - The decision of the government No 315/2014 Coll.		- Act CLXVI of 2012 on the identification, designation, and protection of vital systems - National Directorate General of Disaster management, the agency responsible for the CI protection		- National Critical infrastructure protection program (NCIPP) by the Polish government in 2013	
Legislation or policy requires an annual cybersecurity audit	-No legislation or policy - Only report cover cybersecurity on Slovak's information systems		- No legislation or policy		Act L of 2013 on the electronic information security of central and local government agencies		- No legislation or policy	
Legislation or policy requires the classification of data	- The Act of 11 March 2004 on the protection of classified Information		- The Act 412 on the Protection of classified information 2005		- The Act CLV 2009 on the protection of classified data		- The Act of 5 August 2010 on Protection of classified information	
Legislation or policy requires a chief information officer or chief security officer	- No legislation/ policy - National Security Authority -responsible for information security		No legislation/ policy		Section 17 of Act L 2013 on Electronic information security of central and local government agencies		- No legislation or policy	

	Slovakia	Czech Republic	Hungary	Poland
Operational entities				
Computer Emergency response team (CERT) or computer security incident response team (CSIRT)	- CSIRT.SK established in 2009	- CSIRT.CZ established in 2011 - GovCERT in 2014	- CERT-Hungary established in 2013	- CERT.GOV.PL established in 2008 - CERT Polska in 1996
The national competent authority of network and information security (NIS)	-National security authority for NIS. - Information society section of the ministry of Finance – develop and adopt information security standards	National Security Authority manages the national cybersecurity center (NCSC) under the decision of the government of Czech. - The operation of NCSC is the cooperation between GovCERT and CSIRT.CZ	- National Security Authority for NIS. - The NCSC- operating with the special service for national security.	- CERT.GOV.PL for incident reporting, public education programs, and government but not as wider network and information security authority
Incident reporting platform for gathering cybersecurity incident data	CSIRT.SK – managing the information about cybersecurity incidents - Online reporting structure for recording the incidents	CSIRT.CZ –responsible for incident reporting management The Act on cybersecurity 2014 needs the NSA to manage the incident records	-CERT-Hungary – responsible for incident reporting and collect information about cybersecurity incidents	CERT.GOV.PL - in charge of reporting and responding functions and education programs and consult the government on cybersecurity issues.
National cybersecurity exercises conducted	- Joining in multinational cybersecurity exercises by European Union	- Joining in multinational cybersecurity exercises by European Union	- Joining in multinational cybersecurity exercises by NATO	- Joining in cybersecurity exercises by both NATO and European Union
Public and Private partnership				
Public-private partnership for cybersecurity	No defined public-private partnership for cybersecurity	No defined public-private partnership for cybersecurity	No defined public-private partnership for cybersecurity. However, the NCSC is tasked with private sector for purposes of promoting information and develop long-term cyber strategies	No defined public-private partnership for cybersecurity

	Slovakia	Czech Republic	Hungary	Poland
Industry organize or industry cybersecurity councils	The IT Associate Slovenia (ITAS) for Slovak and international IT companies	No special industry-led platform for cybersecurity	No special industry-led platform for cybersecurity but only Hungarian association of IT companies	Two chambers of commerce: Chamber of commerce for electronics and telecommunications and Chamber of IT and Telecommunications
New public-private partnership	No new public-private partnership	No new public-private partnership but the need to cooperate with the private sector is a key principle for the period 2011-2015	No new public-private partnership	No new public-private partnership
Sector-specific security plan				
Public-private sector plan that addresses cybersecurity	No sector specific joint public-private plans	No sector specific joint public-private plans	Act L of 2013 on the electronic information security of central and local government agencies- providing consideration for sectoral incident management centers.	No sector specific joint public-private plans
Sector-specific security priorities	Not defined yet	Not defined yet	Not defined yet	Not defined yet
Education				
Education strategy to enhance cybersecurity knowledge	- Developing a lifelong learning scheme for IT specialists from the state and private sector - Classes taught at secondary schools - Publishing literature and methodology documents with issues of Information security	- Increasing the cyber and information security awareness of citizens by disseminating relevant information with media - Cooperating with the private sector for training programs on cyber and information security -Integrate cyber and information security at all levels of education	- Integrating cybersecurity in the syllabus of primary, secondary and higher education, training courses for government officials and in professional training courses.	- There is a set of principles on education and training, and a commitment to establish ICT security at higher education sector as a permanent topic. - Using mass media for cybersecurity campaign at young people

Regarding the data from [Table 2.2], Slovakia, Czech Republic, and Hungary had the national cybersecurity strategy while Poland only had a cyberspace protection policy. Besides, we could see that Slovakia was the first country which applied national cybersecurity strategy in the Visegrád group in comparison with the others. Although Poland didn't have a national cybersecurity strategy like the others in the group, Poland was also a pioneer in building the Computer Emergency response team in 2008. Furthermore, V4 is quite similar in several parts such as joining in multinational exercises by EU and NATO, no public-private partnership for cybersecurity, no new public-private partnership, no defined sector-specific security priorities, and focusing on education strategy for the citizens to enhance the cybersecurity knowledge.

Security threats of V4

Visegrád countries 'security environment faces too many security threats for their national security are listed by:

- The weakening of the cooperative security mechanism and of political and international legal commitments in the area of security
- Instability and regional conflicts in and around the Euro-Atlantic area
- Threats from terrorism.
- The proliferation of weapons of mass destruction and their means of delivery
- Cyber-attacks or cyber threats
- Negative aspects of international migration
- Extremism and growth of interethnic and social tensions
- Organized crime, namely serious economic and financial crime, corruption, human trafficking, and drug-related crime
- Threats to the operation of critical infrastructure
- Interruptions of supplies of strategic raw materials or energy
- Disasters of natural and anthropogenic origin and other emergencies

2.6. THE CZECH REPUBLIC

Czech Republic's cybersecurity strategy first established in 2011 and updated version in 2021, it mainly focuses on several essential factors such as principles of security policy, security interests, security environment, and strategy for promoting the security interests [148] [149][Figure 2.2].

Figure 2.2: The Czech Republic cybersecurity strategies' factors

First of all, in the principles of the security policy, it declared the basic concepts, tools, and methods to protect the security of citizens, the state, and how to defend against cyber-threats. Moreover, in this part, the security strategy defined the responsibility of safeguarding the cybersecurity belonged to the local and regional government [148] with the cooperation of Czech citizens, companies, businessmen, and public authorities in order to protect the country's sovereignty and territorial integrity and reduce the cyber risks. However, because of the natural security challenges, with the supporting in cybersecurity strategies in 2015, these security policies not only focus on security concerns but also need to have a set of coherent tools with institutionally and physically cooperation [149].

An important key factor to strongly enhance the defense of the Czech Republic's security is the stability of the EU's economy and politics. Regarding the openness of Czech's economy, especially in market access and energy provides, it supported to development of the Czech's mutually beneficial economic cooperation within international organizations. Czech's principle security mainly focus on staying away with armed confliction and use diplomatic methods with the framework of the United Nation charter to solve the security issues to safeguard the citizens and country. Besides, regarding the membership of NATO and EU, Czech's principles take the benefits of collecting the defense from the NATO system and transatlantic connection for their defense and security.

Secondly, the Czech's security interests are separated into 3 types such as vital interests, strategic interests, and other important interests. In the vital interests, they included the protection of country sovereignty, territorial integrity, political independence, and all other laws to safeguard citizen's rights. Moreover, in the strategic interest's part, there are five main key factors such as supporting, preventing, developing, safeguarding, and maintaining in order to safeguard and promote the vital interests. These are on the table below [Table 2.3]:

Table 2.3: The Czech Republic's strategic interests

Key factors	Mission
Supporting	Democracy, fundamental freedoms, and the legislation Internationally stability via the cooperation with alliance countries Regional cooperation
Preventing	Security threats influenced the Czech's security and its partners Local and regional conflictions and reducing their effects
Developing	The role of OSCE for preventing armed conflictions, democratization, and building mutual trust and security A strategic partnership between NATO and the EU The cooperation in the complementary development of defense and security capabilities The cohesion and efficiency of NATO and EU, and transnational connection
Maintaining	The UN's globally stabilizing role and enhancing the efficiency Functioning and transparent current arms control regime in Europe Security and stability in the Euro Atlantic area
Safeguarding	Internal security and securing the population Economic security and promoting the economy's competitiveness Energy, raw material and food security; and a suitable level of strategic reserves

Additionally, the promoting of other important interests' part enhances the vital, strategic interests and society's resilience towards security threats.

Other important interests:

Beside the strategic interests, the other important interests play an essential role in contributing to the protection of vital and strategic interests; and enhance society's resilience against cyber threats. These other important interests are following by:

- Reducing crime (especially on economic, organized, and information crime) and counteracting the corruption
- Strengthening the Czech Republic's counter-intelligence and defense intelligence
- Promoting a tolerant civil society and preventing the extremism

- Building government institutions and the judiciary more efficient and more professional; enhancing the cooperation between public administration authorities with citizens, and legal entities with individuals or business
- Encouraging the security involvement of civic associations and non-governmental organizations
- Developing public awareness in citizens, and engaging the involvement of the general public in providing for the security
- Promoting the research in science and technology, especially on new technologies with a high added value of innovation
- Developing technical and technological capabilities for the classified and sensitive information's processing and transmission, especially in information protection and accessibility
- Safeguarding the environment.

Thirdly, the increasing security trends including internal and external security threats are more complicated because they are nearly transparent and they are hard to safeguard defense and security.

Threat concerns: military attack directly to the territory of the Czech Republic is low. The decline of security and stability in the EU's flank regions and neighborhood, NATO, and the EU member states can cause the threats. To eliminate these risks, the Czech Republic must be members of NATO and EU; and have good relations with neighboring countries.

The main source of threats: hardline attitudes to fundamental values of society, threatening the concept of the democratic rule of law, and denying fundamental human rights and freedoms. Another source of threat is the power-seeking aspirations of some states that refuse to respect the basic principles of international law, international order. Moreover, the Czech Republic also has the same security threats as the other nations in the Visegrád group. Therefore, the Czech Republic government built several tools to promote security interests not only at a national level but also multilateral and bilateral relations. As a result, they focused mainly on four strategies as follow:

- Collective dimension for protecting security and defense
- The strategy of avoiding and suppression of security threats
- The economic framework for protecting security interests
- The institutional framework for safeguarding the security

In short, the Czech Republic built its strong framework for national cybersecurity for not only the government but also for civil resilience. By clarified the security policy concepts; the security interests; and the security environment, the Czech government listed the factors which can influence directly to national cybersecurity. As a result, they had the general view of the whole security context, then the government could propose a suitable cybersecurity framework at the governance and civil level.

2.7. POLAND

The first national cybersecurity strategy of the Republic of Poland was adopted in 2013. There were six main key factors such as prerequisites and assumptions of the cyberspace protection policy; conditions and problems of the cyberspace; main lines of action; implementation and delivery mechanism of the provision; financing; and assessment of the effectiveness of the policy [Figure 2.3] [150].

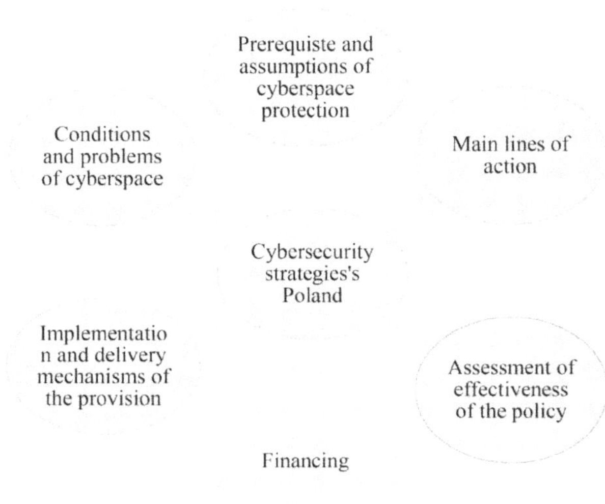

Figure 2.3: Poland national cybersecurity strategy

Firstly, in order to face the ICT security concerns in the cyberspace, Polish government declared the main prerequisites and assumptions of cyberspace as a protection policy to safeguard the security information assets of the State and citizens. This part defined the terms, concepts, security incidents, and organizations that are related to information in cyberspace. Moreover, it included strategic objectives, specific objectives, addresses or extent of the impact, responsibility for the security in the cyberspace protection policy, and compliance of this policy. Secondly, the Polish government also identified the measures – called "ensuring the correctness and continuity of the functioning of the ICT system, facilities and installations" [150] to deploy the essential responsibilities of the State to the citizens and internal security. In addition, it can reduce the potential damage from cyber-attacks of cyberspace and protect the security of the critical infrastructure of the State. Thirdly, to facilitate in implementing this policy, the Polish government suggested some major lines of action, followed by:

- *Risk assessment*: involving general information on types of risks, vulnerabilities, threats, and the responsibilities of each sector or organization deal with them.
- *The security of government administration portals*: guaranteeing the availability, integrity, and confidentiality of data during transferring between government and citizens via e-society or websites.
- *Concepts of legislative actions:* creating the regulations for further actions in applying the provisions of the policy, and enhancing the consideration of the security not only government institutions but all the users in the cyberspace based on the existing regulations.
- *Concepts procedural and organizational actions:* developing the function of the cyberspace Republic of Poland (CRP) through applying the best practices and standards; for example, the management of CRP, the safety management systems in government unit, and the role of representative for cyberspace security.
- *Concepts of education, training, and awareness-raising in security aspect:* improving the education and training for users and creating possibilities of applying the policy. For instance: training for the representative of cyberspace security, introduction the ICT security topics at higher education institutions as a fundamental element, training the secretariat staffs in the government administration, and social public education (children and youth, parents and teachers).
- *Principles of technical actions:* deploying several specific programs to reduce the risk of threats for CRP performance like research programs, creating ICT security incident response team at government level, building the early warning system and maintenance of protecting solutions, testing level of security, and development security teams.

Fourthly, the indispensable part of this policy is the implementation and delivery mechanism of the provisions of the document [Table 2.4]

Table 2.4: CRP's tasks and its responsibilities [150]

Tasks	Responsibilities
Managing and coordinating of the implementation	Council of Ministers - responsible for the information.
Building the national response team for computer security incidents	Three levels: -Level 1: the minister - responsible for information -Level 2: the governmental computer security incident response team (CERT.GOV.PL) with a departmental center for security management of ICT networks and services - responsible for handling computer incidents. -Level 3: administrators - responsible for individual ICT systems in cyberspace
Information exchanging system	An efficient system of coordination based on applicable law and the Act of 29 August 1997 and the Act of 5 August 2010 for exchanging information between government, military, civilian, and international cooperation.
Methodology and forms of cooperation	Developing the forms of cooperation between the authorities responsible for security and fighting against computer crime. Decrease delays in computer incident response
Cooperation with organizers	Cooperation with some sectors such as communication, ICT networks, finances, transportation, providers in energy, energy resources, and fuel. Coordinating with ICT device, systems factories, and telecommunication organizers
International cooperation	Expanding the cooperation between government agencies, public organizers, representatives, and non-governmental institution to enhance the security of CRP and international security

Fifthly, in order to implement the policy, it requires the costs for executing the tasks; however, the cost of starting the tasks should be estimated and be decided by the results of the risk assessment in specific projects. Every organization needs to indicate the tasks with a clear explanation related to cybersecurity and estimate the cost for the tasks as well. Besides, the essential expenses part will be limited to the budgetary consumption in the budget act for each year. Last but not least, the Polish government created several measures to evaluate the effectiveness of the policy such as effective standard, products standard, result standard, impact standard. Moreover, in this policy, it also clarified that one important element to examine the effectiveness of actions is creating the scope of the tasks for each individual and identify the responsibility of their exercise; then, regarding on the report of the progress to monitor the effectiveness of actions. In addition, an obligation thing is that the users need to announce immediately the computer incidents to an administrator or suitable CERT in order to take actions and handle it to restore an acceptable level of security in case of data or system security breaches.

In summary, the Republic of Poland step by step built the complete structure of national cybersecurity strategy from the conditions, prerequisites, and problems of cyberspace to the actions, responsibilities of each department, and financial support for doing the tasks and provision for the future. It makes their policy more effective with the supporting adequate response, evaluation of computer security incidents and improves the recovery process to an acceptable level of security. It leads Poland to be a leader in cybersecurity role in Visegrád countries.

2.8. HUNGARY

The national cybersecurity strategy of Hungary (NCSS) was established in 2013. It focused on a unique model of cooperation between state and non-state actors. Moreover, it is based on the standards of EU and NATO cybersecurity concepts and followed the current cyber security strategies (values, environment, objectives, tasks, and tools)[151] [152]. In addition, the Electronic Information Security of Central and Local Government Agencies established the first legal framework for almost all Hungarian cybersecurity organizations in Act L of 2013. Regarding this law, Hungarian government organizations and bodies approached information security at different levels. These levels are based on the tasks, the importance, and the requirements of the organizations, individuals, measures, and documents. In order to deploy the cybersecurity strategies, the Hungarian government identified the cybersecurity organizational structure. The main structure of Hungary national cybersecurity strategy based on four factors such as political and strategic management, national and international cyber policy coordination; operational cybersecurity capabilities, cyber incident management, and coordination; military cyber defense; and crisis prevention and crisis management; [152] [Figure 2.4].

Figure 2.4: Hungary cybersecurity strategy structure

Hungarian government put five major objectives to build the strong cybersecurity in strategy as follows [153], [152], [Table 2.5]:

Table 2.5: Hungary national cybersecurity objectives

Objective	Mission
Creating response capability	-Preventing, detecting, managing, and correcting malicious cyber activities, threats, attacks or emergencies, information leakage. -Establishing GovCERT-Hungary
Building a secure environment	- Providing protection for national data assets, functions of vital systems and facilities - Building an efficient, fast and loss-minimizing correction system in emergency case
Enhancing education	- Training and research development for international best practices - Declaring special role for National University of Public service - Ensuring the standard of cybersecurity education
Safeguarding the future generation	- Establishing secure cyberspace for children and future generations with international best practice
Implementing international standards	- Guaranteeing the quality of IT, communication products, and services - Applying international security certification standards

Firstly, political coordination and management play an essential role in the national digital economy and information infrastructure development, especially in cybersecurity. Cybersecurity requires not only national cooperation but also international coordination because there is no fence to avoid cyber threats. The National Cybersecurity strategy created the National CyberSecurity Coordination Council as the highest political coordination body of Hungary with the responsibility for cybersecurity issues. This organization includes several ministerial leaders and public entities; for example, State Secretaries of Defense, Interior, Foreign Affairs And Trade, Finance, National Development, Hungarian National Bank, and the National Media and Telecommunication Authority. Moreover, this Council works with Cybersecurity working groups (Homeland Security, Child Protection, and E-Government) as well as senior experts. Besides, this Council includes the National Cybersecurity Forum to create a chance for business CEOs, academic and Non-Governmental Organizations (NGOs) to meet the governmental decision-makers. In this forum, national or international companies can share cybersecurity knowledge and experiences together. The Ministry of Interior manages the operation of the National Cyber Security Coordination Council, the National Security Authority, the National Electronic Information Security Authority, and GovCERT-Hungary in order to enhance the safety of classified information and electronic systems or sensitive data; handle the data of central and local government agencies; and mitigate with incident handling process.

Secondly, the operational cybersecurity capabilities and cyber incident management mainly focus on the governmental computer emergency response team in Hungary. GovCERT-Hungary was founded in 2013 with approximately 4000 institutions as partners, and it was supervised by the Ministry of Interior. It offers the services for Hungarian governmental administration like backbone system, critical infrastructure, and the municipalities. In addition, it cooperates with the private sector for enhancing the information exchanges, increasing awareness in the network security, creating cooperation with the international CSIRT and critical information infrastructure protection (CIIP) community. This organization aims to create dynamic malware analysis, higher event correlations, remote exams in order to make online malware and knowledge database for a cyber-alert early warning system [151]. In order to guarantee cooperation, make the task execution and incident handling processes more effective, the National Cyber Defense Institute was established in 2015. It has a connection with the National Electronic Information Security Authority, the Cyber Defense Management Authority, the Military Computer Incident Response Capability (MilCIRC), and several international forums such as Forum of Incident Response and Security Team, International Watch and Warning network, and the European Government CERTs group. Likewise, GovCERT-Hungary also joins in national and international cyber defense and crisis management exercises regularly. Thirdly, the Hungary National Military strategy in 2012 defined Hungary cyberspace as the fifth important domain because the Hungarian Defense Forces (HDF) not just fight the physical dimension but also from cyberspace as well. Then, in 2013 the Ministry of Defense published the cyber defense concept of the Hungarian Defense Forces. It declared the general requirement for the cybersecurity tasks for HDF and their organizations. Furthermore, it built the legal, regulatory environment; increased the security awareness and knowledge, R&D, cooperation with stakeholders; and promoted cyber defense capabilities of HDF in three levels such as initial, basic and full cyber defense capabilities. In 2014, the Minister of Defense established the Military National Security Service to develop the Computer Incident Response Capability (MilCIRC) and Military Computer Emergency Response Team (MilCERT). The Ministry of Defense is only responsible for safeguarding military communication in peacetime, while the Communications, Information Systems and Information Security Directory of the Hungarian Defense Forces General Staff manage the security concerns of military networks, governmental organizations, authorities, and other partners like NATO and EU cooperation. Hungary participated in NATO Cooperative Cyber Defense Centre of Excellence (NATO CCD COE) as a sponsoring nation in June 2010 with the purpose to improve the cyber defense capability; cooperation and information sharing with NATO; and contribute to the education, R&D, consultation for CCD COE. Finally, the Ministry of Interior created the National Directorate General for Disaster Management (NDGDM) for managing three main pillars: fire protection, civil protection, and industrial safety. It helps to prevent disasters, control the protection activities, reduce the negative effects of emergencies, and enhance reconstruction and recovery. It also plays an important role in the protection of cybersecurity and critical infrastructure as well. Besides, within the framework of the national inspectorate general of industrial safety of NDGDM, the Critical Infrastructure Cyber Incident Response Center was founded with the main aims to guarantee the network security of critical infrastructure factors, mitigate industrial security incidents, train and join in industrial and network security exercises. Last but not least, this center also cooperates with GovCERT-Hungary in solving industrial security incidents.

In summary, the Hungarian government cybersecurity strategy is based on the standards of EU and NATO cybersecurity concepts and the control of the Ministry of Interior. This strategy is a combination of State and non-State actors, military and law enforcement, and economic and political stakeholders in order to build the free and secure use of cyberspace for users. Additionally, the Hungarian government strengthened several organizations to deal with cybersecurity incidents (GovCERT, MilCERT, MilCIRC, and HDF) to safeguard cyberspace and create a secure digital environment.

2.9. SLOVAKIA

Regarding the strategic interest in economic development and the global cooperation of the EU, Slovakia's national cybersecurity strategy focused on three major purposes such as prevention, readiness, and sustainability. As a result, they can help to protect Slovakian digital space from security incidents, guarantee the response and mitigate ability towards security incidents and recover the operation after the incident, and keep or improve Slovakia's competence in information security, respectively [154]. Moreover, the Slovakian government clarified seven key main functions in national cybersecurity strategy, followed by:

- Safeguarding of human rights and freedoms: using all measures to make Slovak digital space and personal data secure
- Developing awareness and competence in information security: enhancing the education activities and culture of using ICT through several projects by the Ministry of education, science, research, and Sport to improve the security awareness and competence of ICT users.
- Creation of a secure environment: related to building a legal framework depended on basic rights and freedoms as well as the clarification of the responsibilities and competencies for the public administration and coordinating standardization
- Improving the effectiveness in information security management: creating the information sharing and warning system for threats detection and response to incidents, integrating the CSIRT.SK into the Europe cooperation (ENISA, European Public-Private Partnership for Resilience - EP3R)
- Ensuring sufficient protection of the critical information infrastructure: improving the information security in state agencies; and applying new secure products, systems, and conditions to ensure the security for national critical infrastructure
- National and international cooperation: enhancing international cooperation depended on national requirements and priorities.
- Improving national competence: analyzing the information security quality and possibilities for education and training; recommending a training system; building up research and development; and providing economic competitiveness.

Additionally, the Slovakian government made a clear structure for national cybersecurity and cyber defense with 5 essential parts such as political and strategic level cyber security management, cyber incident management and coordination, military cyber defense, intelligence, and cyber aspects of crisis management [Figure 2.5], [155].

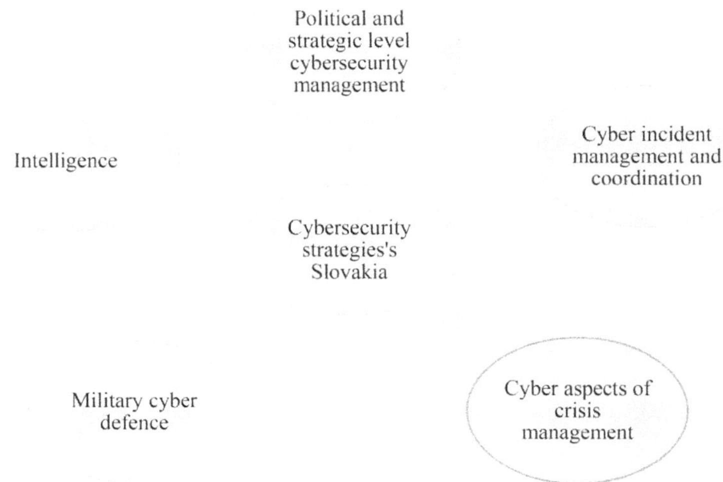

Figure 2.5: Slovakia cybersecurity strategy structure

Firstly, in the political and strategic level cybersecurity management part, the Slovak government clarified the differences between the management and information security of top-secret and unclassified information to a well-organized structure for cybersecurity and cyber defense itself. Ministry of Finance and National Security Authority (NSA) are responsible for creating legislation, standards; and protection of classified information, cryptographic services, respectively. Moreover, the NSA offers protection for foreign classified information shared with Slovakia based on international agreement and cooperates with the other NSA of other members and security authorities of international organizations. Secondly, the Slovak government created the national computer security incident response team (CSIRT) for dealing with cybersecurity threats and risks. This organization works independently and is supported by the Ministry of Finance. It has three departments (technical, national information and Communication infrastructure, and educational department) with the responsibility for collecting the information about cybersecurity threats; incident handling; and implement education concepts for managers, IT staff, public institutions, and for every individual. CSIRT also provides both reactive and proactive services, or public institutions, Commercial Corporation, organizations, and individuals such as alerting security threats or vulnerabilities, investigating incidents or malware, responding to incidents, education, giving information, configuration and infrastructure maintenance, and building awareness in information security. Furthermore, although CSIRT is the only official organization registered in Slovakia, there are several other organizations such as the Sanet (Slovak academic network, member of TERENA), ISACA Slovak chapter, ITAS (IT Association of Slovakia), Sasib (Slovak Association for Information Security), and Slovak is also a member of Central and Eastern European Networking Association (CEENet) – with the major purpose in academic, research and education in computer network security cooperation. Thirdly, the Ministry of Defense (MOD) created cybersecurity for the military (CSIRT.MIL.SL) in order to monitor, evaluate, and measure the information security aspect. This organization is also responsible for enhancing the awareness of cybersecurity via education, supporting the Computer incident response capability, and creating defense toward cyber attacks. This team also cooperates with foreign CSIRTs and other international organizations, however, it lacks qualified individuals. Besides, cybersecurity is the most part is exercised by the ministry of defense under two levels: under the Department of CIS and support section, and the General staff of the armed forces. This part not only took part in installing, maintaining, securing classified information, managing cryptographic hardware and software for the Ministry's information system but also safeguarding the registry of documents from NATO and EU. In addition, the CSIRT team is aimed to have three major groups such as analytics-technology, prevention, reaction, research, and special studies group to combat the cyber-attacks. Fourthly, the Slovak Information Service is a central intelligence and security service organization that can safeguard the intelligence protection of the Slovak Republic. This organization is under the control of the Government and the Security Council and it helps to collect the Intelligence and Open Source Intelligence (OSINT) and share the information with other law- enforcement for EU platforms and NATO structure. Last but not least, the Slovak government established Act No. 45/2011 on the critical infrastructure and declared the responsibility of the Ministry of Interior and other Ministries with sector or sub-sectors [Table 2.6]. This leads the information security coordinator or owner of the infrastructure to deploy the security plan and improve the technology in order to secure the critical infrastructure feature.

Table 2.6: Cyber aspects of crisis management [155]

Sector	Subsector	Organization
ICT	Information systems and networks, Internet	Ministry of Finance, CSIRT.SK
Electronic Communication	Satellite communication, networks and stable and mobile services of electronic communications	Ministry of transport, construction and regional development
Transport	Road, air, water, rail	Ministry of transport, construction, and regional development
Post	Post services, a system of payments and procurement activities	Ministry of transport, construction and regional development
Health		Ministry of health
Energy	Electricity, gas, crude oil, mining	Ministry of economy
Water and Atmosphere	Drinking water, water construction, meteorology	Ministry of economy
Industry	Pharmaceutical, chemical, metallurgical	ME Slovak Republic

In supporting the national cybersecurity strategy 2009, the Slovak government defined the strategic purposes, several solutions, and legal framework [Figure 2.6] for cybersecurity in the new cybersecurity strategy of Slovakia in 2015 – 2020, followed by [156]:

Strategic purposes:

- Safeguarding national cyberspace - a system operating conceptually in a coordinated manner, efficiently, effectively, and on a legal basis
- Increasing the security awareness of all components of society
- The private and academic sectors, as well as a civil society, actively participate in the formulation and implementation of the policy of the Slovak Republic in the area of cyber-security.
- Providing for both national and international levels in collaboration efficiently.
- Adopting the measures and respecting the protection of privacy and basic human rights and freedom.

Solutions:

- Creating an institutional framework for cybersecurity administration
- Building and adopting a legal framework for cybersecurity

- Identifying and deploying basic mechanism for securing the administration of cyberspace
- Providing, developing, and proposing a system of education in the area of cybersecurity
- Specifying and implementing a risk control culture and a system of communication between the stakeholders
- Making active international collaboration
- Strengthening science and research in the area of cybersecurity.

Furthermore, this document offers the formulation of regulations, standards, methodology, rules, security policies, and other necessary tools to support the cybersecurity of the Slovak government.

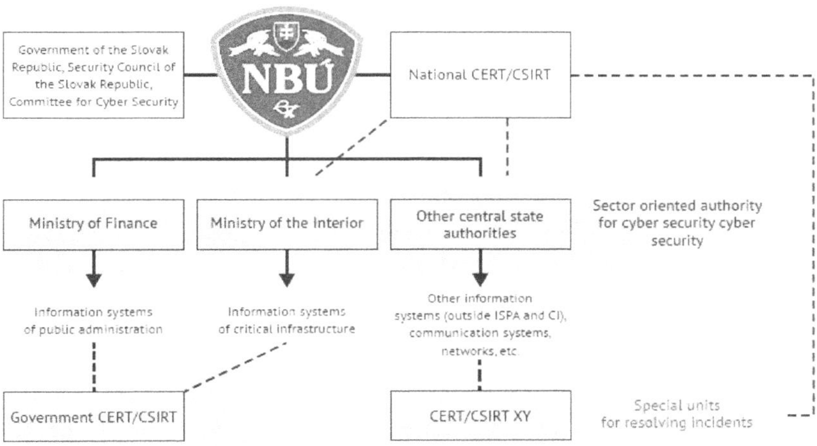

Figure 2.6: Propose a framework structure for managing cybersecurity for the Slovak government [156]

In short, the Slovakian government noticed that the area of cybersecurity plays a crucial part in using information and communication technology. Therefore, they built a strong collaboration between public administration (CSIRT and CERT) and the private or academic sector; legal framework, basic mechanisms to evaluate cyber threats, and computer incidents to ensure cyberspace. Likewise, they also focus on implementing the education system to spread knowledge and increase awareness of cybersecurity areas from many levels such as primary, secondary, university, and experts.

Key findings for Europe cybersecurity

ENISA

In 2004, the European Parliament and the Council established the first cybersecurity agency for the EU – the European network and Information

Security Agency (ENISA). Its body has three major elements such as The Management Board, The Executive Director, and The Permanent Stakeholder's Group. The main purposes of this agency are enhancing the capability of the Member States to prevent or respond towards network information security issues, improving a high level of expertise, providing assistance or advice to the Commission and the Member States, updating and boosting Community legislation in network information security [157]. This organization also created general CERT for all Member States (CERT-EU) and a part of CSIRT based on the Directive on security of Network and Information Systems (NIS Directive).

NIS Directive

European countries also have the official cybersecurity strategy " The Open, Safe and Secure Cyberspace" which was formed in February 2013 [157], [158]. In this general cybersecurity, it mainly focuses on five priority strategies, following by:

- Accomplishing the cyber resilience
- Extremely diminishing cybercrime
- Promoting cyber defense policy and capabilities to the Common security and defense policy (CSDP)
- Boosting the industrial and technological resources for cybersecurity
- Setting up an international cyberspace policy for the EU and improve core EU values.

In addition, this strategy also clarified the roles and responsibilities of many actors such as CERTs, law enforcement, NIS competent authorities at both national and EU-level [Figure 2.7] in dealing with cybersecurity incidents. It also expressed the guidelines of the EU's support in major cybersecurity attacks or incidents on EU governments, businesses, and individuals.

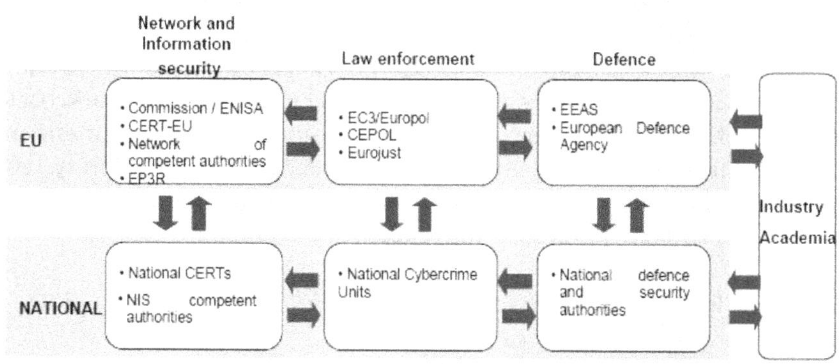

Figure 2.7: Different legal framework operation at national and EU-level [158].

GDPR

The GDPR is a new regulation for EU countries that is effected in May 2018 with the main purpose to handle data for all organizations [159]. Moreover, it also gives guidance for the security of data processing within 99 articles [160]. Particularly, Article 32 of GDPR established the requirements for Data controllers and Data Processors in deploying technical and organizational tools for guaranteeing a level of data security during data processing [161], as follow:

- *"The pseudo and encryption of personal data;*
- *The ability to ensure the ongoing confidentiality, integrity, availability, and resilience of processing systems and services;*
- *The ability to restore the availability and access to personal data in a timely manner in the event of a physical or technical incident;*
- *A process for regularly testing, assessing and evaluating the effectiveness of technical and organizational measures for ensuring the security of the processing"*

Regarding this article, the organizations in EU nations can get fatal financial consequences if they failed in data security (up to 2% of their annual global sales or 10 million euros). As a result, the implementation of GDPR not only helps the EU nations to protect their data during processing and transferring amongst them, ensure the data security but also safeguards the organizations avoid financial penalty.

NIST 800 Revision 53

National Institute of Standards and Technology (NIST) – (a non-regulatory agency of the U.S. Commerce Department) is responsible for creating information security standards, guidelines for federal information systems including federal agencies, state, local, private sector organizations, and tribal governments under the Federal Information Security Modernization Act (FISMA) in 2002 [162], [163]. In addition, it also supports agencies to develop suitable security policies and controls to secure all federal information systems. It built the cybersecurity framework in order to help organizations recognize the cybersecurity risks and know how to mitigate the damage from these risks and respond to cybersecurity incidents via customized measures. NIST published a Cybersecurity Framework (CSF) including standards, guidelines, and best practices to control cybersecurity issues [164]. In 2017, the NIST established the fifth special publication "SP" 800-53 with the aim of indicating these regulations can be used for all organizations and all systems not just federal organizations and information systems [165]. Currently, North America and Europe's organizations are using the NIST frameworks like NIST 800-53, the CSF, and the newly updated NIST Risk Management Framework (RMF). Especially, the NIST SP 800-53 contains many recommendations which meet the requirements under Article 32 of GDPR, therefore, it can be used for any organizations in both North America and EU members.

Contractual Public-Private Partnership (CPPP)

CPPP is a part of the EU cybersecurity strategy. It was established in 2016 by the EU commission and the EU cybersecurity organization [166]. This partnership aimed to enhance the cooperation between the public and private sectors at the beginning state of the research and innovation process. Moreover, it also helps to promote the cybersecurity industry and supports critical infrastructure operators and research institutes to develop cybersecurity solutions such as energy, health, transport, and finance. CPPP is based on the funding from the H2020 project (the biggest EU research and Innovation program with approximately 80 billion euros during 2014 to 2020 for creating a genuine single market in knowledge, research and innovation to secure the EU Member States) [167]. At the beginning state, there were three initiative research Public-Private Partnerships such as Factories of the Future (FoF), Energy-efficient Buildings (EeB), and Green Cars (EGVI in Horizon 2020) but now, it has seven more cPPPs in industrial sectors and technology areas like 5G, Sustainable Process Industry (SPIRE), Robotics, Photonics, High-Performance Computing (HPC), Big data, and Cybersecurity [168]. As a result, CPPP plays an important role in industrial development roadmaps for the EU at national and regional levels.

Digital Single Market Initiative

Digital single market is a policy of the EU single market which includes digital marketing, e-commerce, and telecommunication. It is part of the Digital Agenda for Europe 2020 program and it was established in 2015 by the European Commission [169]. This strategy created digital opportunities for people and businesses in the digital environment. Besides, it promotes the EU's position as a leader in the digital economy over the world. The main purposes of a digital single market are as follows [170]:

- Building the digital single market
- Promoting the European digital industry
- Creating a European data economy
- Enhancing connectivity and access
- Supporting funds in network technology
- Boosting in digital science and infrastructures
- Building a digital society
- Improving trust and security
- Promoting media and digital culture

Three Seas Initiative

A political and economic inter-governmental platform between the Adriatic, the Baltic, and the Black Seas – The Three Seas Initiative (3SI) was established in 2015 to develop the integration of Central and Eastern European countries (CEE) and improve their position in the EU [171], [172]. This includes 12 European Members States: Austria, Bulgaria, Croatia, the Czech Republic, Estonia, Hungary, Latvia, Lithuania, Poland, Slovakia, and Slovenia. This initiative firstly aimed to enhance cybersecurity in three areas: energy, infrastructure, and digital. Then, this organization contributes to improve cohesion and unity within the EU Member States via several activities such as joining cross-border projects, developing popular security models and standard for 5G, implementing the free flow of non-personal data privacy, developing Industry 4.0, securing e-commerce centers, fighting information warfare, creating digital innovation hubs or competence centers and developing cybersecurity policies. Lastly, this initiative's purpose is to strengthen transatlantic ties.

The North Atlantic Treaty Organization (NATO)

NATO created a National Cybersecurity Strategy (NCS) framework which included three main pillars such as authorization, dimensions, and difficulties [173]. The authorization has five elements that require the management of the incident cycle; for instance, cyber diplomacy & Internet governance, critical infrastructure & crisis management, intelligence & counter-intelligence, cyber military, and fighting cybercrime. Besides, there are three dimensions which are different stakeholder groups like "government, national actors, and international – transnational groups". However, NATO also clarified five difficulties which member nations should balance between the costs and influences on the freedom, economic development, and NCS requirements, following by:

- Encouraging the economy vs enhancing national security
- Modernizing infrastructure vs protecting critical infrastructure
- Private sector vs public one
- Protecting data vs sharing information
- Freedom of expression vs political stability

NATO also pointed out that the NCS strategy might not be applied as a unique model for every country. Therefore, it depends on how a nation concentrates on cyber difficulties and takes them into consideration at government levels.

European Public-Private Partnership for Resilience (E3PR)

European Public-Private Partnership for Resilience was founded in 2009 on Critical Information Infrastructure Protection (CIIP). This partnership's purpose firstly maintained cross-border cooperation for all EU members (27 countries) with four major pillars [174]:

- Encouraging information sharing and stock-taking of good policy and industrial practices to promote popular understanding
- Discussing public policy priorities, aims and measures
- Offering standard requirements for the security and resilience in the EU
- Identifying and developing the adoption of good standard practices for security and resilience

Then, this cooperation engaged the public and private sector to collaborate in a multilateral, open and conference for partnership and agreement to achieve new five pillars for security, follow by:

- Preparing and preventing
- Detecting and responding
- Mitigating and recovering
- International cooperation
- Criteria for EU's critical infrastructure in the ICT sector

Key findings for V4 cybersecurity cooperation
Why V4 cooperation is good?

The V4 cooperation showed that it created a friendly relationship in international politics. This relationship regards the common history, shared geographical neighborhood, economic collaboration, and awareness of popular interests [175]. With the V4 cooperation, it can contribute to promoting not only EU and NATO in security structure but also in cyber defense more effective, functional and powerful based on their similar interests. Furthermore, regarding the cooperation of state, government, and administrative authorities, it may support V4 face to social, cultural, and security challenges and ensure their position in the same region. In fact, the immigration crisis is one of the important security aspects that requires the cooperation of V4 to work together with the EU in supporting the admission mechanism. Additionally, regarding V4 cooperation, it can help V4 in solving the energy problems because they depend on importing energy issues and they are lack of integrated energy market, infrastructure, and interruption in supplying energy resources. Moreover, with similar cyber threats, V4 cooperation can promote military capabilities and cooperation in the armed forces via sharing military exercises, combat capabilities, and defense experiences. For example, Poland creates cyber-attacks capacity in the army. The Czech Republic is strong not only in technical but also in cybersecurity. Hungary is good at engineering training. Slovakia is leadership in the public sector in cybersecurity [176].

Cooperation in cybersecurity in V4
Similar

- Joining in Digital Three Seas Initiative cooperation for economic growth, development IoT, Artificial Intelligence (AI), 5G, digital infrastructure, tactical cooperation against cyber threats and disinformation [177].
- Hungary and Slovakia's cybersecurity strategy belong to the Ministry of Interior and they have civil resilient cooperation.
- The internal cybersecurity of Visegrád countries have the offense capability by law or regulations
- They set up the CPP cooperation and strong cooperation with the University

Different

- Poland and the Czech Republic have strong CERT but Slovakia and Hungary are still immature of CERT to defense against the cyber-attacks.
- Hungary and national cybersecurity institutions focus on civilian law capabilities and it belongs to the Ministry of Interior and civilian security services. Besides, Slovakia and Czech Republic cybersecurity belong to the Ministry of Interior while Poland has cybersecurity capabilities belongs to the Ministry of Military.

Therefore, Hungary and Polish cyber center organizations cannot cooperate because of the former on the Interior side and the latter on the Military side. Moreover, the Czech Republic is different from the three countries because it has offense capabilities by law.

2.10. CONCLUSION

This chapter has briefly described the foundation of Visegrád countries (history, the purpose of cooperation, and its mechanism). It also presented how V4 countries had cooperation within V4, with EU and other international organizations (NATO, Western Balkans, Benelux group, OECD, and UN). This chapter also examined the similarities and differences of the V4 cybersecurity strategies themselves [178]. For example, the difference between the Czech Republic cybersecurity strategy and the others is that it mainly focuses on the essential role of information security and its loss; emphasizes the cybersecurity's awareness-raising of public and private sectors; combines cybersecurity with protecting human rights and democratic states' standards.

While Hungary's cybersecurity strategy involves the implementation of the rules of national interests within the State and global context, ensuring the close relationship between government, academia, business sector, and civil society depending on their shared responsibilities. In addition, Poland cybersecurity strategy ensures the State's safety in cyberspace; promotes the cooperation of proactive activities from State and private sectors with other entities in the energy, transport, telecommunication, and health sectors; establish suitable standards and good practices to support private or non-private organizations (institutions, research organizations, scientific and NGOs) in cybersecurity risk management. Similar to the Czech Republic, Slovakia cybersecurity focuses on awareness-raising in political, legal, economic, social, and technical organizations to provide safe cyberspace. Likewise, Slovak clarified that education was also the key factor for cybersecurity, as a consequence, the collaboration of the public and private sector, academic organizations, and civil society is the highlight in their strategy. Rather, the Slovak government pointed out cybersecurity as a key component of national security and it needs to follow the legislation in foreign policy, defense and civil emergency planning, and intelligence services in EU and NATO documents. Last but not least, the author points out several organizations for Europe cybersecurity cooperation and legal frameworks such as ENISA, NATO, the Three Seas Initiative, and E3PR; Digital Single Market Initiative, NIS directive, GDPR, NIST 800-53, and CPPP, respectively. Regarding the V4 cooperation advantages listed above, V4 cooperation expressed that these countries are considered as one nation with a great impact on EU and NATO in enhancing cybersecurity, cyber defense, and several challenges like immigration issues and energy. In another way, "Cybersecurity in Visegrád countries shares similarities regarding goals, strategies, and strength to align with European Union Member States regarding armed forces, cybersecurity, and national security", was formulated.

Chapter Three:

POLICIES, STRATEGIES, COOPERATION IN ASIAN COUNTRIES

This chapter is aimed to investigate the policies, strategies, and cybersecurity cooperation in Asian countries. In particular, the European cybersecurity strategies and Asian cybersecurity strategies are considered. In addition, it was intended to discover security capacity, problems, policies, and legal frameworks of each Asian and ASEAN country based on the collecting of information from media communication, the official and national documents or publications regarding legal framework and strategies development, and valuable statistical data from Asian cyber wellness profiles. Besides, a suitable cybersecurity cooperation model is developed in order to apply for Asian countries by analyzing the V4 cybersecurity cooperation model.

3.1. GENERAL POLICIES, STRATEGIES, COOPERATION OF ASIAN COUNTRIES

Cyber risks impact

According to Internet World Stats 2017, the Internet users in Asia accounted for approximately half of Internet users worldwide [Figure 3.1], [Figure 3.2]. However, they are still immature with cybersecurity, exercises, or cooperation to counter cyber incidents, or cyber-attacks. As a result, it is a honey pot for hackers to abuse such drawbacks. In fact, in 2016, hackers attacked some ASIAN countries by withdrawing US$81 million from the Bangladesh Central bank, accessing and leaking details of 3.2 million customer cards from several Indian banks, stealing US$65 million of bitcoins from Hong Kong-based digital currency exchange Bifinex, using malware to steal US$2.17 million from eight banks in Taiwan. In 2017, a remarkable attack in Korea was recorded, indicating that seven main banks were threatened by a distributed denial of service attacks claiming for ransom payment [179].

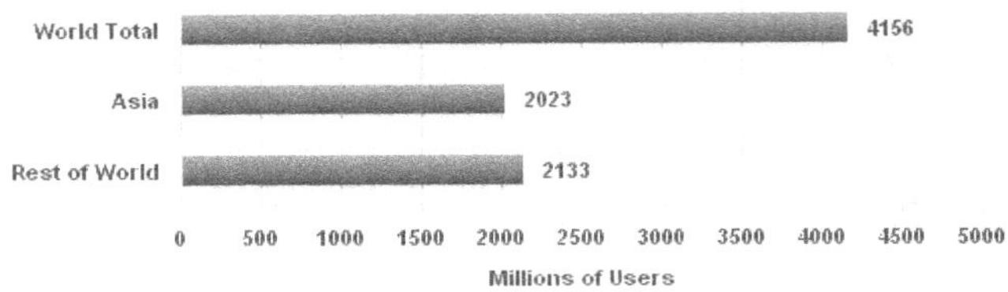

Figure 3.1: Internet users in Asia in 2017 [180]

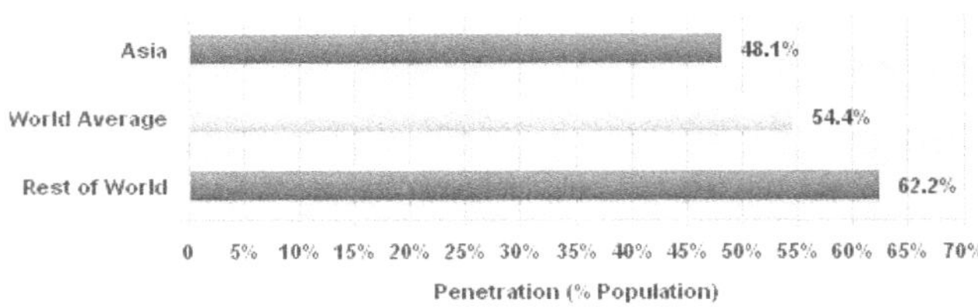

Figure 3.2: Internet penetration in Asia in 2017 [180]

The legal framework in the cybersecurity of Asian countries

In Europe, every country has a proper cybersecurity strategy but all of them have to comply with the foundation of the European Union laws and regulations. However, Asian countries mainly focus on economic growth and cybersecurity cooperation in trading, e-commerce. Some of them pay attention to building a cybersecurity strategy to protect their national interests and civilians [Table 3.1].

Asia Pacific Computer Emergency Response Team (APCERT)

Asia is an organization to support, provide safe, clean, and reliable cyberspace for the Asia Pacific region through global cooperation. It has 30 teams from 21 economic countries in Asia. This organization networks trusted computer security experts in the Asia Pacific area to enhance cybersecurity awareness, competency towards computer security issues or cyber-attacks. Furthermore, this organization mainly targets in several missions, following by [181]:

- Improving Asia Pacific area and international cooperation on information security
- Developing the measures to mitigate local and global network security incidents
- Providing information sharing and technology exchange between its members such as information security, computer virus, vulnerabilities, and the like
- Boosting collaborative research and development on subjects of members' interests
- Supporting inputs or recommendations to solve legal issues about information security and emergency response over regional boundaries.

Table 3.1: Legal framework of some Asian countries in cybersecurity

	Legal foundation	Data protection	Operation entities	Public-Private Partnership	Sector specific cybersecurity plans	Education	Additional Cyber-law indicators
China	-No national cybersecurity strategy - Only several government policies with advice on cybersecurity - No specific law on cybersecurity -State secrets law 2010	Cybersecurity law in 2017	-National CERT, CNCERT in 2002 -National information security-belong to different government bodies -Little public information about their operations and objectives	- A little activity in public-private partnership	-No joint public-private sector plan	-No national cybersecurity education strategy - Only some ad hoc education initiatives by the CERT and ministry of industry and information technology	-Imposing a range of legal and policy restrictions on cybersecurity service providers
Hong Kong	-No national cybersecurity strategy	The Personal Data (Privacy) Ordinance (Cap. 486 of the Laws of Hong Kong) (Ordinance) in 1996 [182], [183]	Hong Kong CERT (HKCERT), the cybersecurity and technology crime bureau (CSTCB), the office of the Privacy Commissioner For Personal Data (PCPD), the Hong Kong Monetary Authority (HKMA), the Hong Kong Institute of Directors, the Office of the Government Chief Information Officer (OGCIO), Hong Kong Internet Exchange (HKIX), Hong Kong Internet Registration Corporation Limited (HKIRC), Hong Kong Internet Service Providers Association (HKISPA), Hong Kong Police Force (HKPF), and the Office of the Communications Authority (OFCA), Government Information Security Incident Response Office (GIRO)	Internet Infrastructure Liaison Group, the OGCIO EGCCSS, Working Group on Cloud Computing Interoperability Standards (WGCCIS), Working Group on Cloud Security and Privacy (WGCSP) and Working Group on Provision and Use of Cloud Services (WGPUCS).	-No joint public-private sector plan	-No national cybersecurity education strategy	Online child protection

	Legal foundation	Data protection	Operation entities	Public-Private Partnership	Sector specific cybersecurity plans	Education	Additional Cyber-law indicators
Japan	-Cybersecurity strategy in 2013 -Basic Law on cybersecurity 2014 -New state secrets law in 2013 for making stronger security practices on solving sensitive information and stronger penalties in case of unauthorized access	The Act on the Protection of Personal Information ("APPI") and the Personal Information Protection Commission ("PPC") in 2007 [184]	-National CERT, JCERT/CC in 1996 -Cybersecurity strategy Headquarters in 2014 under Basic law on cybersecurity	- A mature public-private partnership structure including J-CSIP	- No joint public-private sector plan	-Cybersecurity strategy in 2013 includes detail and comprehensive commitment to educating young people on cybersecurity	-Avoiding undue legal and regulatory restrictions on cybersecurity service providers
South Korea	-National security and defense focusing on cybersecurity -Cybersecurity Master plan in 2011 but more cyber-defense strategy - Minor gaps in their legal framework	The law on Personal Information Protection Act, "PIPA" in 2011 [185]	-Both KrCERT/CC and KNCERT(only government) - Korea Internet and Security Agency responsible for information security	-KrCERT/CC liaise with the private sector as a part of incident response duties -No formal public-private partnership for cyber or information security	-No joint public-private sector plan	-Korea Information security agency- responsible for users' internet usages, and the agency conducts online and broadcast awareness raising campaigns	- Undue restrictions on cybersecurity service providers
North Korea	No national cybersecurity strategy	No information	- The National Cyber-Security Center -The Korea Internet & Security Agency (KISA) - The National Police Agency's Cyber Terror Response Center	No information	No joint public and private sector plan	No information	No information

	Legal foundation	Data protection	Operation entities	Public-Private Partnership	Sector specific cybersecurity plans	Education	Additional Cyber-law indicators
Singapore	-National Cyber security masterplan in 2013 - Cybersecurity Agency of Singapore 2015 -National Cybercrime action plan 201	The Personal Data Protection Act of 2012 [186]	-SingCERT in 1997 - Infocomm Development Authority- responsible for information communications policy, including cybersecuri-ty	-Singapore government agencies -working closely with the private sector in cybersecurity aspect - A formal commitment to the development of public-private partnership	-The Infocomm Security Masterplan 2 in 2008 -developing sector-specific security programs, particularly CI. - Gov-Tech agency – responsible for development Cybersecurity for public and government	In national cybersecurity masterplan in 2018 -including a strong commitment to cybersecurity education	- Avoiding undue legal and regulatory restrictions on cybersecurity service providers
Malaysia	-No single cybersecurity strategy -Having the collection of policies and strategies as Malaysia's cybersecurity policy.	The Personal Data Protection Act 2010 (PDPA) enacted in 2013 [187]	-National CERT (MyCERT), cyber999 as the chief authority on information security	-Organizing an award event which doubles as an annual convention on cybersecurity in a public-private partnership model	-Public-private sector - the main key to identify security concerns and 10 critical sectors for cybersecurity	- The CyberSafe program -offering a comprehensive suite of materials and activities relating to cybersecurity	-Restrictions on global cybersecurity providers -Avoiding undue legal and regulatory burdens
The Philippines	National cybersecu-rity strategy in 2005	The Data Privacy Act of 2012 [188]	the National Computer Emergency Response Team (NCERT), Cybercrime Investigation And Coordina-tion Center (CICC), CSP-CERT, CSIRT, the Philippine National Police (PNP), National Bureau of Investigation (NBI), Department of Justice (DOJ), government CERT(GCERT)	- No government and public sector agencies	- No joint public-private sector plan	No information	- Online child protection, cybercrime Act and Criminal code

	Legal foundation	Data protection	Operation entities	Public-Private Partnership	Sector specific cybersecurity plans	Education	Additional Cyber-law indicators
Indonesia	-National cybersecurity strategy -Weak legal framework -No clear classified security law or policy and security practices -No specific cybersecurity provisions	No general law on data protection	ID.SIRTII/CC, National CERT, ID.CERT	-No dedicated cybersecurity public-private partnership - The CERT as the main liaison body for the private sector	-No joint public-private sector plan	-No cybersecurity education strategy	-Discriminatory procurement preferences, local testing requirements, and a limit on data flows
Thailand	- No national cybersecurity strategy	No general law on data protection	Official and legally government CSIRT (ThaiCERT)	No any official recognized national or sector-specific programs for sharing cybersecurity assets within public and private sector	Ministry of Information and communication technology (MICT) – responsible for national cybersecurity strategy, policy, and roadmap	MICT as national and sector-specific for education, training program in raising awareness cybersecurity	Specific legislation on child protection (Thailand Penal code, computer crime act
Laos	No national cybersecurity strategy or policy	No general law on data protection	LaoCERT, CIRT in 2011	No any officially recognized national or sector-specific programs for sharing cybersecurity assets within the public and private sector	-No joint public and private sector plan	-No cybersecurity education strategy	- Article 138 of the criminal code, Article 86 of the law on the protection of the rights and interests of children

	Legal foundation	Data protection	Operation entities	Public-Private Partnership	Sector specific cybersecurity plans	Education	Additional Cyber-law indicators
Cambodia	No national cybersecurity strategy	No general law on data protection	National CamCERT in 2011, CSIRTs, ISP/IX	No dedicated cybersecurity public-private partnership	-No joint public-private sector plan	No national or sector-specific educational or professional training for raising awareness cybersecurity	Specific legislation on child protection Convention on the rights of the child
Vietnam	No national cybersecurity strategy National Anti-crime 2012-2015 A draft law on information security	No general law on data protection	-VNCERT in 2005. - Other operational entities are limited.	-Not defined public-private partnership -VNCERT liaises closely with the private sector	-No joint public-private sector plan	No general public awareness or education strategy	Setting certain procurement restrictions and technology authority on cybersecurity service providers

3.2 KEY FINDINGS FOR ASEAN CYBERSECURITY

Problems

There are four main problems in order to enhance cybersecurity in ASEAN such as political, economic, social, and miscellaneous problems. Firstly, there is non-state cooperation in politics amongst ASEAN nations, therefore, it is difficult to solve the cyber-attacks when they happened. Secondly, the difference in the economic status of each nation in the same region is a big gap to develop cybersecurity capacity in order to mitigate cyber-threats. Thirdly, cyber-threats or cyber-attacks can influence social life and national stability. Hackers can use their skills to penetrate government databases and make the citizens lose trust in their government. It can lead to the destruction of the social and moral fabric of a nation like the series of attacks by "Anonymous" in Singapore in 2013 [189]. Finally, because of the boom of technology, hackers have become more sophisticated in their attacks. This increases challenges when attackers aim to the less digitally developed countries with fewer experts or technology to deal with.

Political problems

The policy of ASEAN countries is not interferential; therefore, it interrupts the development of cybersecurity. When the attack happens, countries cannot help others immediately because of fear of violating this policy. Hackers may use this advantage for their attack. Moreover, there are different perceptions and opinions about cybercrime, therefore, the main focus and attention of ASEAN countries not on cybersecurity. In fact, according to Hein [190], ASEAN countries responded to cybercrime quite low and fragmented because some of them haven't had experiences in serious cyber threats and they haven't recognized cyber security's importance. Furthermore, they lack efficient strategies to counter cyber threats or cyber-attacks. Indeed, among ASEAN nations, there is no common organization or system to enhance cybersecurity. In addition, less digitally developed countries (Vietnam, Laos, Myanmar, and Cambodia) haven't got any solutions or they hesitate to make decisions regarding threats or attacks; therefore, these are the serious issues to counter cyber threats. Furthermore, there is an absence of common governance or legal framework at the ASEAN level which challenges cybercrime [191]. Almost all ASEAN governments and organizations are lack trust and transparency in sharing incident information or threat intelligence, as a result, it is hard to investigate, prevent, and mitigate cyber-attacks. This weak point may lead to limit mandates to share specific cyber incident information across intelligence agencies.

Economic issues

In general, almost all private companies mainly focus on economic benefits with new innovation features from new technology to attract consumers to buy or use these technologies but they rarely concentrate on protecting their users [192]. Hence, most hackers may steal or gain illegal access to sensitive and financial information of users, government agencies, or economic organizations for making attacks. Besides, there is a delay time in identifying cyber-attacks after it happened. It can lead to adverse effects. Likewise, because of the differences in research development, sector and digital literacy are also a gap between ASEAN members. Each member's economic status is relevant to its level of digital development. Some developed countries with high economic status can invest more in the research and development sector while less developed members have difficulties in doing that due to high cost. This makes it difficult in exchanging technologies among countries because such technology will only be suitable for some developed countries. However, to narrow this gap, in 2011 there was a master plan for the ASEAN Cyber University. This project was established by the Ministry of education of the Republic of Korea with the purpose to improve higher education in the ASEAN region, lessen the gap among ASEAN member states and support ASEAN's efforts for regional integration [193], [194].

Social problems

Cyber threats or attacks may have strong and negative impacts on the development of a country or they can destroy the infrastructure [195]. Besides, hackers usually work with terrorist organizations because of their capabilities and financial resource. Hackers may seek manpower or use their technical skills to exploit government databases in order to destroy the cyber defense of a nation. This makes citizens lose their trust in their government and scared to live in an unstable country, as a result of the destruction of social and country's moral fabric [195].

Miscellaneous problems

Nowadays, cyber hackers use more complicated methods for their attacks. It is more difficult to mitigate the damage and recovery of stolen data or sensitive information, especially with less digitally developed countries as well as lack of experts and technology.

Cooperation in Computer Security Incident Response Team

There are several cybersecurity organizations to support, improve cooperation, response, and information sharing among the Computer Security Incident Teams (CSIRTs) in economies of the Asian Pacific regions. For instance, firstly, the Asia Pacific Computer Emergency Response Team (APCERT) was founded in 2003. It functions as a forum for CSIRTs and CERTs in the same region. It has 30 operational members from 21 countries in Asia (Australia, Bangladesh, Brunei, Bhutan, People's Republic of China, Taiwan, Hong Kong, India, Indonesia, Japan, South Korea, Laos, Macao, Malaysia, Mongolia, Myanmar, New Zealand, Singapore, Sri Lanka, Thailand, and Vietnam) [196]. Besides, it has two categories of members: operational and supporting members. The former members are dealing with the function of CSIRT/ CERT on full-time as a leading or national CSIRT/CERT within their own economy. The latter members are cybersecurity entity which can contribute to APCERT operations and CSIRT/CERT functions. This organization creates policies, practices, and procedures for enhancing the Asia Pacific regional and international cooperation on information security, facilitating information and technology sharing, improving the collaborative research and development on subjects of members' interest, raising awareness on computer security incident response, and supporting other CERTs/ CSIRTs for effective computer emergency response [197]. Secondly, the organization of Islamic Cooperation – Computer Emergency Response Team (OIC-CERT) has a similar mission to APCERT. Its members are from 23 countries (Azerbaijan, Bangladesh, Brunei, Côte D'Ivoire, Egypt, Indonesia, Iran, Jordan, Kazakhstan, Kuwait, Libya, Malaysia, Morocco, Nigeria, Oman, Pakistan, Qatar, Saudi Arabia, Sudan, Syrian Arab Republic, Tunisia, United Arab Emirates, and Uzbekistan) [198]. It creates a platform for increasing cybersecurity capabilities, developing cooperation initiatives and possible partnerships to fight against cyber threats by leveraging global collaboration.

Strong cybersecurity capacity nations

In Asia, there are several countries with strong cybersecurity capacity such as China, Hong Kong, Japan, South Korea, North Korea, Singapore, and Malaysia. They built their national cybersecurity strategy or cybersecurity policy, as well as legal framework, cyber laws, cybersecurity capacity, cyber defense, and governance organizations to deal with cyber-threats. Moreover, they have good international cooperation with different countries in the same region and others in order to share knowledge, best practices and increase cybersecurity awareness towards cyber-attacks.

China

China first established its national cybersecurity strategy in 2015 with several objectives, as noted below [199]:

Objectives

- Peace: management arms races in cyberspace, conflictions, and other threats to international peace.
- Security: controlling cybersecurity risks, protecting national cybersecurity systems, securing core technologies and equipment and network information systems.
- Openness: sharing information technology standards, policies, and markets; exchanging technology, attack on cyber terrorism and cybercrime; completing multilateral, democratic and transparent international Internet governance system
- Order: protecting the public's right, human rights, right to express ideas, participate, and other lawful rights and interests in cyberspace.

Concepts

- Respecting and protecting sovereignty in cyberspace
- Peaceful use of cyberspace
- Governing cyberspace according to the law
- Comprehensively manage cybersecurity and development.

The Chinese government declared several important tasks in order to protect China's cybersecurity, national interests' sovereignty, security and develop cyberspace [Table 3.2]. Moreover, it also helps to improve humanity, peaceful use, and common governance cyberspace.

Table 3.2: Chinese strategic tasks for cybersecurity [199]

Tasks	Purpose
Defending sovereignty in cyberspace	-Managing online activities based on constitution, laws, and regulations -Applying several measures: economic, scientific, technological, legal, diplomatic, military, and administrative measures to protect the country's sovereignty in cyberspace -Fighting against the destruction and damage activities towards the country's sovereignty via a network.
Protecting national security	- Preventing separatism, treason, rebellion, or subversion to people's democratic, regime, and destruction activities. - Protecting activities to steal or leak State secrets
Protecting critical information infrastructure.	- Using all measures to protect critical information infrastructure, important data from attack and destruction -Controlling and applying laws, regulation, rules, standards to safeguard critical information - Establishing a cybersecurity mechanism for information sharing between government, sectors, and enterprises
Strengthening the construction of online culture	- Developing a positive and upward online culture - Encouraging the expansion of new businesses - Improving the protection of minors online - Boosting online civilization construction, online theory
Attacking cyber terrorism, law-breaking, and crime.	-Developing online anti-terrorism, espionage, fraud, theft, drug trafficking, arms, and anti-theft capabilities - Control at the online source and lawful prevention
Perfect network governance systems	- Managing and governing the website in a lawful - Completing cybersecurity law and regulation systems - Increasing the construction of a network governance system with legal norms, technological protection - Encouraging social organizations to participate in network governance - Building and completing national cybersecurity technology for system - Applying cybersecurity projects, academies and improving education in cybersecurity

Tasks	Purpose
Enhancing cyberspace protection capabilities	- Building cybersecurity protection forces - Developing cybersecurity defense means - Discovering cyber intrusions - Creating a backup force for protecting national cybersecurity
Strengthening international cooperation in cyberspace.	- Boosting international cyberspace dialogue and cooperation - Reforming the global internet governance system - Strengthening bilateral and multilateral cybersecurity cooperation in global, and regional organizations - Supporting developing countries and backward regions - Establishing a multilateral, democratic, and transparent international Internet governance system -Building a peaceful, secure, open, cooperative, and orderly cyberspace

In order to remain the communist party's ruling power, the Chinese government has three highest-level decision-makers to create cybersecurity policies for cyberspace such as the Politburo Standing Committee, the State Council, and the Central Military Commission [200]. Furthermore, it also has several agencies like the Ministry of Industry and Information Technology (MIIT), the Ministry of Public Security (MPS), and the Ministry of State Security (MSS). Firstly, the MITT was founded in 2008 to centralize information technology development. This part is responsible for all State Council on managing information, setting standards, practicing exercises, investigating network security to respond to cyber-attacks, vulnerability databases, malicious IP, and so on as well as the National Computer Network Emergency Response Technical Team/ Coordination Center of China (CNCERT). The MIIT also creates guidelines, policies, laws, and regulations for the State Administration for Science, Technology, and Industry for National Defense (SASTIND). Besides, the Commission for Science, Technology, and Industry for National Defense (COSTIND) also has similar tasks with the MIIT. Secondly, the MPS is in charge of investigating cybercrime and protect critical infrastructure with research labs. This Ministry also controls the commercial products and commercial information security companies, especially the Great Firewall of China.

Thirdly, the MSS focuses on countering espionage, foreign intelligence, and domestic intelligence, separatism, terrorism, and religious extremism. In addition, there are several other government research institutions to support three ministries in order to safeguard the critical information, for example, the State Encryption Bureau (managing the import and export of any encrypted devices for a party, civilian, military encryption), the Chinese Institute of Contemporary International Relations, The Chinese Academy of Engineering, The Chinese Academy of Sciences, the Cyber Security Association of China. They help to speed up the development of industry standards and coordinate research on cybersecurity. Last but not least, the Chinese government established the cybersecurity law in 2017 as the first national-level law for cybersecurity and data privacy protection to address the framework and regulations in order to ensure national security and public interests [201].

International cooperation

China provides the UN and the UN Security Council with counter back cybercrimes and cyber terrorism. Besides, China improves regional cooperation on ICT within the framework of the Asia Pacific meeting [202]. Moreover, China expands its cooperation partnership with several organizations of the international community; for example, it takes part in some international conferences, forums like China-Japan-Korea cyber policy consultation, ASEAN Regional Forum (ARF), Boao Forum for Asia, the Conference On Interaction And Confidence Building Measures In Asia (CICA), Forum on China-Africa Cooperation (FOCAC), China-Arab States Cooperation Forum, Forum of China And The Community of Latin American and the Caribbean States and Asian- African Legal Consultative Organization [202]; as a result, it can enhance the cooperative initiative in the digital economy, dialogue exchange on cyberspace with other regional groups.

In brief, China is a power and the world's most populous country in East Asia. Besides, China is an early technology approach nation not only in the nuclear weapon but also in cybersecurity; therefore, they built a national cybersecurity strategy to protect national cyberspace and critical infrastructure; enhance their cyberspace protection capabilities; develop international cooperation in cyberspace; and counter against cybercrimes and cyber-terrorism.

Hong Kong

The Hong Kong Productivity Council established and managed The Hong Kong Computer Emergency Response Team Coordination Center (HKCERT) in 2001 in order to improve awareness of malicious software (ransomware or malware) attacks. This organization cooperated with computer security incident response for local enterprises, Internet users and supported in exchanging of information with other CERTs. Moreover, it was considered as a point of contact on cross-border security incidents [203]. HKCERT also joined in FIRST to share technical information, tools, best practices, security vulnerabilities, cybersecurity attacks, and incidents of computer systems and networks with other members to handle and promote the prevention programs. Moreover, there are several organizations, institutions, and agencies in order to support the Hong Kong government in cybersecurity; for instance, The Cyber Security And Technology Crime Bureau (CSTCB) in 2015, the Cyber Security Information Portal (CSIP) in early 2015, the office of the Privacy Commissioner For Personal Data (PCPD), the Hong Kong Monetary Authority (HKMA), the Hong Kong Institute of Directors [203], the Office of the Government Chief Information Officer (OGCIO), Hong Kong Internet Exchange (HKIX), Hong Kong Internet Registration Corporation Limited (HKIRC), Hong Kong Internet Service Providers Association (HKISPA), Hong Kong Police Force (HKPF), and the Office of the Communications Authority (OFCA) [204]. Furthermore, Hong Kong was the first nation in Asia which established personal data privacy legislation and privacy regulator by the Personal Data Privacy Ordinance (PDPO) [182], [183]. This law covered both public and private sectors in the data privacy regulatory framework and it focused on marketing regulation, international data transfer, cybersecurity, data breaches, and implementation of the use of personal data [205]. Last but not least, the Hong Kong government built the organizational framework for handling security incident response at the government level, known as Government Information Security Incident Response Office (GIRO) [206]. This organization is responsible for coordinating and providing the operation of the Information Security Incident Response Team (ISIRT) of the bureau and departments (B/Ds).

International cooperation

In order to share cybersecurity assets and best practices with other nations, Hong Kong has several international ties with organizations such as ITU, Interpol, APCERT, and Asia-Pacific Economic Cooperation (APEC) [207]. In particular, Hong Kong is an operational member of APCERT, joined the APCERT Drill with the topic" Emergence of a New Distributed Denial Service Threat" in 2017 [208] and it has a good relationship with other CERTs like JPCERT/CC and CNCERT [209].

Regarding the OGCIO, it helps the Hong Kong government create information technology strategies, programs, and countermeasures toward cybercrime, cyber-attacks. Moreover, it supports IT services and government to maintain Hong Kong's position as Asia's leading digital city. Furthermore, with international cooperation with other organizations, Hong Kong can share innovation technology, best practices, experts, and education to ensure cybersecurity for its national security, critical infrastructure, and citizens.

Japan

Japan's government recognized the importance of information security in 2001. Therefore, the Japanese government first started the e-japan strategy in 2001 with the purpose to become the most developed countries in ICT by developing the infrastructure for ICT. For example, they invested in building high-speed internet access (approximately 30 million households)[210]. Then, they built up the ICT infrastructure for e-government and e-commerce in 2003. In fact, there were 96% electric filing of Government of Japan (GOJ) and 23% Internet trade of all exchange by the end of 2003 [210]. Moreover, in 2004, the Japanese government launched the Policy roundtable for realizing a ubiquitous network society in order to make networks that can be used anytime and anywhere. From that period, the Japanese government mainly focused on investing in cybersecurity to make Japan the most developed country in ICT's world. For instance, the first cyber security strategy was launched in the periods 2006 to 2008, the second national strategy in 2009 to 2011, information security strategy to protect the nation in 2010, and cybersecurity strategy 2013 [Figure 3.3], [211]. In 2014, the Japanese government delivered new cybersecurity principles to agencies. This policy helps to protect the critical infrastructure against cyber threats, following by some major policies: supporting and enhancing safety concepts, developing information sharing and incident response, risk control, and improving the standards for Critical Information Infrastructure Protection (CIIP) [212], [213]. In addition, because of the major and serious cybersecurity attacks (Mitsubishi Heavy industry in 2011, Sony Pictures Entertainment in April 2011, and Japan Pension service in 2015) [214], the Japanese government drafted the new cybersecurity strategy in 2015. This new strategy aimed to put the Japanese government, ministries, agencies, and other organizations into high-level concentration towards cybersecurity and created the standard measures to implement new regulations or cyber laws for these organizations.

Besides, the Japanese government set up several organizations, agencies, and cybersecurity centers to manage the information sharing and flow of information, monitor the cyber threats, cyber incidents or cyber-attacks, implement training programs, and technological operations such as: the Information Security Policy Council (ISPC), Center for International Public Policy Studies (CIPPS), Ministry of Economy, Trade and Industry (METI), National Information Security Center (NISC), National Police Agency (NPA), Ministry of Defense (MOD), Japan Computer Emergency Response Center (JPCERT/CC) (1996), Information Technology Agency (IPA), Initiative for Cyber Security Information Sharing Partnership of Japan (J-CSIP), Industrial Cybersecurity Center of Excellence (COE), Network Incident Analysis Center For Tactical Emergency Response (NICTER), the National Cyber Training

Center, Ministry of Internal Affairs and Communications (MIC) [Table 3.3], [215], the Cyber Clean Center (CCC), the Advanced Cyber Threats Response Initiative (ACTIVE) [214], the Information Security Center Council (ISCC) [173], and Control System Security Center (CSSC) [216]. Likewise, the Japanese government also gave the Act on the Protection of Personal Information ("APPI") and the Personal Information Protection Commission ("PPC") - a central agency for supervisor governmental organizations in privacy protection [184]. Regarding this Act, it specifies personal information, sensitive personal information, anonymized information, and guidelines for collecting, processing, and transferring data to the third party for safeguarding the data and strengthening national security.

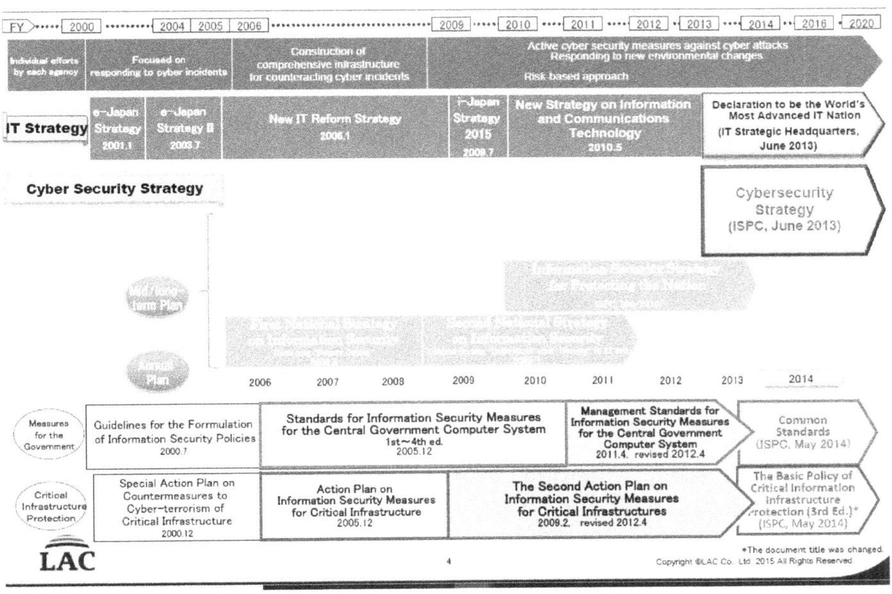

Figure 3.3: History of Japan cybersecurity [211]

Table 3.3: Japan's cybersecurity organizations [215]

Organizations	Functions
The Information Security Policy Council (ISPC) (2011)	- Top-level of the Japanese government's cybersecurity advisory body - Improving the cooperation between public-private sectors
Center for International Public Policy Studies (CIPPS)	- Private sector – creating public policy issues in international affairs and diplomacy issues
Ministry of economy, trade, and industry (METI)	- Establishing the IT policies
National information security center (NISC)	- Coordinating government efforts
National police agency (NPA)	- Fighting the cybercrimes
Ministry of Defense (MOD)	- Establishing national security
Japan Computer Emergency Response Center (JPCERT/CC) (1996)	- First CSIRT in Japan - Cooperating with service providers, security vendors, government agencies, and industry organizations - A member of FIRST and APCERT - Providing computer incident responses - Cooperating with local and global CSIRTs
Information –Technology Agency (IPA)	- Monitoring the next-generation government security operation coordination team for central government - Supporting sharing cyber threat information framework - Establishing a cyber rescue and advice team against the attack of Japan (J-CRAT)
Initiative for cybersecurity information sharing a partnership of Japan (J-CSIP)	- Protecting critical infrastructure companies
Industrial cybersecurity center of excellence (COE)	- Developing the human resources - Evaluating the security and reliability of Industrial control system/ supervisory control and data acquisition (ICS/SCADA) - Researching and analyzing cyber threat intelligence
Network incident analysis center for tactical emergency response (NICTER)	- Monitoring cyberattacks and visualizing them
The national cyber training center	-Offering SecHack365 program to train students under 25 years old -Implementing 100 cyber defense exercise with recurrence exercises for central and local municipal government officials and critical infrastructure personnel

Organizations	Functions
Ministry of internal affairs and communications (MIC)	- Establishing the IoT cybersecurity action program 2017 to improve IoT security and prepare for Tokyo 2020.
The Cyber Clean Center (CCC)	- A honey pot to capture malware and monitor their behaviors - Identifying and warning infected users
The Advanced Cyber Threats Response Initiative (ACTIVE)	- A project – reorganized from CCC to help in the prevention of malware compromising - Alerting Internet users from malware - Reducing the number of malware infection
Control System Security Center (CSSC) (2012)	- A technology research center to enhance and authenticate cybersecurity

International cooperation in cybersecurity

Japan and the USA started to build cooperation in 2013 to strengthen strong security and defense cooperation with the purpose of increasing the capacity-building efforts. Then in 2015, there were a new version "Guidelines for Japan and US defense cooperation. In Southeast Asia, the USA and Japan have common security interests; as a result, their cooperation can guarantee the security and stability of their democracies, counter back terrorism, piracy, and so on [214]. Additionally, Japan and India shared the same vision of free and secure cyberspace and international law in the first and the second meeting in cybersecurity in 2012 and 2017 to improve confidence-building measures or develop investments in cybersecurity cooperation [217]. Likewise, Japan and the EU also join in Japan-EU ICT policy dialogue and other forums or discussions to build a cooperative framework on information security with EU nations like Japan- UK cyber conferences, Japan-EU Internet Security Forum [218]. Moreover, Japan has a good relationship with ASEAN countries. It also cooperates with ASEAN nations in several meetings and projects related to the cybersecurity field; therefore, it helps to improve bilateral cooperation, cybersecurity awareness, share the best practices with each member country by exchanging its views towards cyberspace, information on cybersecurity strategies and discussing the possibility of cooperation, to counter against cyber-attacks.

In summary, the Japanese cybersecurity strategy targets to improve government cybersecurity, critical infrastructure, associated with corporations and academia; enhance cyberspace's awareness for the business and its citizens; and apply countermeasures for cybercrime and cyberspace defense. Regarding the Japanese cybersecurity strategy, the Japanese government built an inter-organizational cooperation framework that enhanced their national cybersecurity capabilities, developed coordination and collaboration among the parties, agencies, and public-private sectors, enabled immediate response to cybersecurity incidents or cyber-attacks. Besides, the Japanese government develops cooperation with international organizations like the USA, the EU, and the ASEAN in order to share best practices, raise cybersecurity awareness, promote technological innovations, and support local and global CSIRTs.

South of Korea

In 2011, the South Korean government established the country's Cybersecurity master plan but it mainly focused on cyber defense strategy than a cybersecurity strategy [219], [220]. This national cybersecurity master plan based on three areas like an investment in security capabilities, building a legal framework, and international cooperation [221]. In addition, there are three cybersecurity agencies to detect, protect, handle cybersecurity problems or cyber-attacks such as the National Cyber-Security Center, the Korea Internet & Security Agency (KISA), and the National Police Agency's Cyber Terror Response Center [222]. Moreover, it also established two computer emergency response teams at the national level such as KrCERT/CC (1996) and KN-CERT (2004). KrCERT works with the private sector and is controlled by KISA while KNCERT deals with the public sector and works as a part of a national cybersecurity center. Furthermore, the Republic of Korea (ROK) military is responsible for both defensive and offensive cyber capabilities to counter North Korea's cyber capabilities. Indeed, the 2014 South Korean Defense White paper reported that North Korea operated about 6000 cyber warfare troops, interruption of military operations, attacks on South Korea's national infrastructure [221]. ROK is known as a wired-connect country; as a result, their cybersecurity or policy issues are based on using traditional media, social media, and academic arguments. To increase cyber awareness, South Korea's government applied several national military service programs, training, or educational programs in universities or hire professional experts from the private sector to build cyber expertise. For example, in 2008, the Ministry of Education created the Educational Cyber Security Center (ECSC) to improve security principles at some research universities. This center cooperated with KrCERT and the National Cybersecurity Center in computer incident responses. South Korea's government built several cybersecurity bodies with different responsibilities such as the Ministry of National Defense, The Ministry of Science, ICT And Future Planning, The National Intelligence Service, the National Police Agency's cyber bureau, the Korea Communications Commission (KCC), and The Ministry of Security and Public Administration to identify, analyze cybersecurity incidents and improve the security level of national information and communications networks. Although South Korea's government has several laws and regulations on information protection (such as laws on Personal Information Protection Act (PIPA), military secrets, telecommunications, cybercrime, e-government, and information infrastructure laws), South Korea hasn't had comprehensive cybersecurity or critical infrastructure legislation. However, they had the Korea Information Security Management System (K-ISMS) as a national cybersecurity standard [223] and the national cybersecurity crisis management framework [224], [Figure 3.4]. In order to share the cybersecurity experiences, assets with other nations, KISA has cooperated with several organizations and other CERTs such as Office Of Cybersecurity And Information Assurance (OCSIA UK), Israel National Cyber Bureau (INCB), Checkpoint Israel, Microsoft, MacAfee, CERT Australia, CERT Romania (CERT-RO), Chinese CERT (CN-CERT), APCERT, FIRST, Japan CERT (JP-CERT) and Cybersecurity Institute (STS) of Kazakhstan [225].

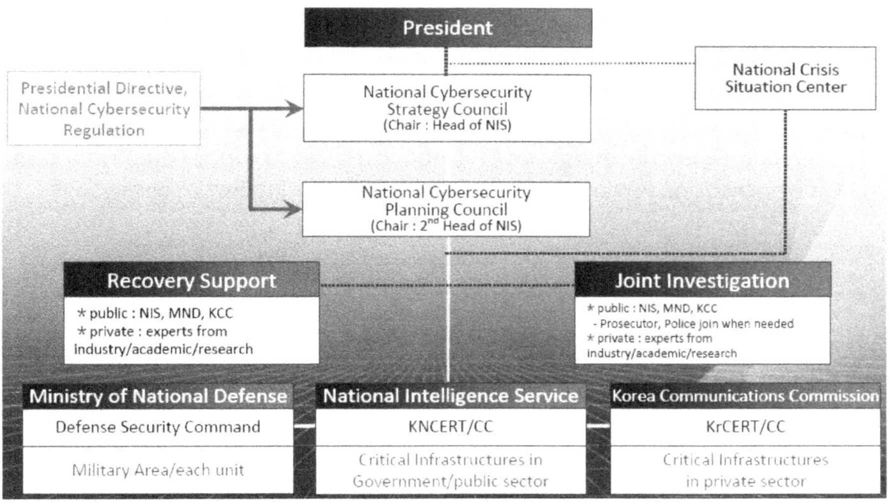

Figure 3.4: South Korea's National cybersecurity crisis management framework [224]

Shortly, although South Korea is a world leader in IT and a leader of the USA and EU in Internet penetration, it got many cyber-attacks, cyber warfare, cybersecurity incidents, or issues from the conflict of North Korea. Therefore, the South Korean government created a national cybersecurity strategy to collaborate with government agencies, public-private sectors, enterprises for making efforts against the cyber-attacks. Moreover, regarding the local international cooperation with many organizations, it can strengthen the security of national critical infrastructure, detect and stop cyber-attacks at the national level and improve its cybersecurity infrastructure.

North Korea

North Korea or the Democratic People's Republic of Korea (DPRK) is considered one of the hardest intelligence targets in the world because it has very little information about its cyber strategy [226]. Recently, North Korea has been related to cyber-attacks in South Korea and the USA. Therefore, North Korea may invest in developing its cyber capabilities for both political and military purposes. DPRK's strategy focused on asymmetric and abnormal operations to deal with the USA and ROK's military strength.

North Korea's cyber operations are a way of targeting vulnerabilities based on cyberspace for national and military activity. In peacetime, cyber capabilities help DPRK to defeat the risk of revenge or operational risk. The organization of DPRK's cyber operations has two main organizations which are responsible for peacetime provocation and wartime disruptive operations such as the General Bureau of Reconnaissance (GBR) and the General Staff Department (GSD) [Figure 3.5]. GBR has been related to cyber-attacks towards South Korea and the USA. In fact, in 1998, there were more than 30000 cyber-attacks against South Korea [227]. While GBR controls DPRK's cyber capabilities, GSD is responsible for military operations and units in supporting traditional military operations towards interruptive attacks and cyber operations.

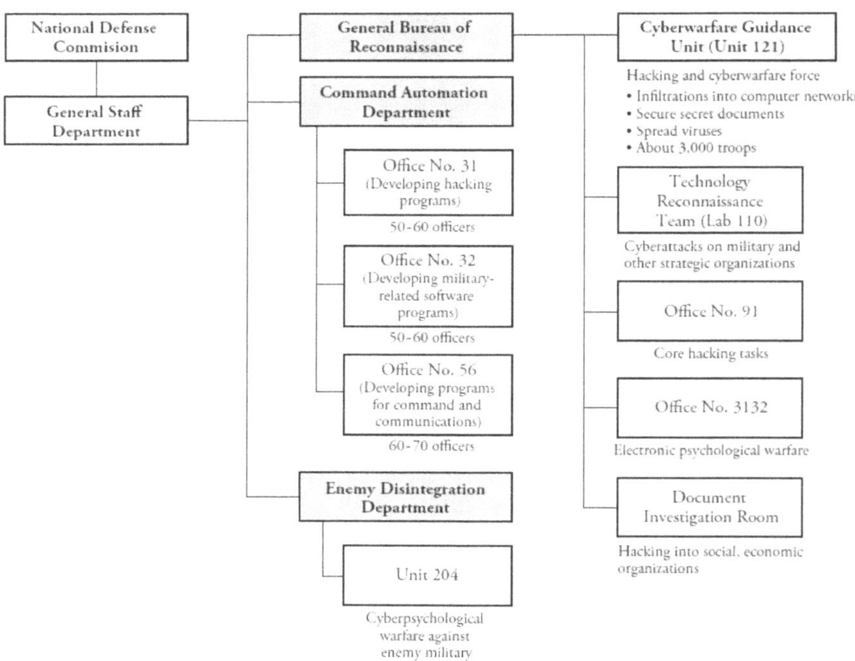

Figure 3.5: North Korea's cyber warfare organizations [227]

Moreover, North Korea mainly considers the defense, interruption, attacks, and analysis of cyberspace [228]. Additionally, the main pillars of Pyongyang's cyber strategy are firewalls, cyber espionage, network attacks, and the distribution of disinformation. Regarding cyber-attacks, it can support DPRK's budget for several reasons like a low cost for a startup the business, anonymity of the users or money transactions, variability of earning money (freelance tasks to illegal targeted attacks), possibility of avoiding UN sanctions (the ban or workers' hiring from DPRK) [229]. As a result, DPRK's cyber strategy regarding cyber capabilities as their offensive and defensive way to protect their national security and get profits in cyberspace.

North Korea is an isolated and disconnected country from the world network together via cyberspace [230]. As a result, this makes North Korea become an extremely secure cyber domain because it virtually turns off the global Internet. Besides, North Korea's cybersecurity strategy majors in developing cyber capabilities both political and military aspect to ensure cybersecurity for national cyberspace. On the other hand, North Korea's strategy develops cyber operations as cyber-attacks, cyber espionage, or cyber sabotage towards South Korea, the USA, and other nations to keep their national position in the region and take financial advantages for the regime.

Singapore

The Singapore government took cyber threats into serious consideration and it started early a few years ago. They built cybersecurity plans to ensure Singapore's digital environment safely, safeguard Singapore against cyber threats, and strengthen public sector cybersecurity capabilities by composing the masterplan and organizations [231], [Figure 3.6].

Moreover, Singapore's cybersecurity strategies in 2016 focused on 4 pillars and their functions such as: building resilient infrastructure, creating safer cyberspace, developing a vibrant cybersecurity ecosystem, and strengthening international partnerships [Table 3.4], [231]. After the remarkable attack in breaching the networks of the National University of Singapore and Nanyang Technological University in April of 2017. It stole government data that contained defense projects, foreign affairs, and transport sector information. As a result, in July 2017, the cybersecurity Bill was announced to the public with the purpose to create a framework for critical information security (CII); set up a framework for sharing of cybersecurity information with and by CSA; and support cybersecurity agency of Singapore (CSA) and control the cybersecurity threats [232], [233]. Singapore government intends to build this country to be a Smart Nation by 2020; therefore, in 2018 Singapore government established a new Omnibus cybersecurity bill with new roles to ensure security incidents will be informed to the government within hours; and security incidents response plan should be in place.

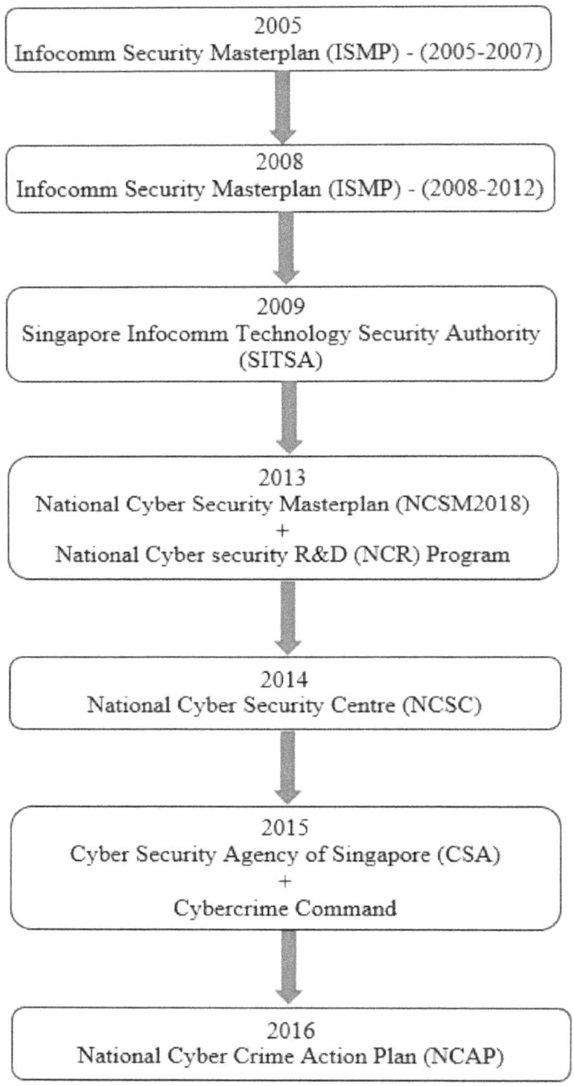

Figure 3.6: List of Singapore's cybersecurity plans during a decade

Table 3.4: Singapore's cybersecurity pillars and its functions

Pillars	Key points
Building a resilient infrastructure	Enhancing CII protection program to create robust and systematic cyber risk management for all critical sectors Combining multi-sector cybersecurity exercises Expanding national resources like the National Cyber Incident Response Team (NCIRT), national cybersecurity Center (NCSC), creating cybersecurity Act for the cyber agency of Singapore (CSA) Increasing the efforts to secure government systems, networks, citizens and official data. Focusing on five sectors such as emergency services, e-government, banking and finance, utilities, and transport and healthcare [234]
Creating safer cyberspace	Creating National Cybercrime Action Plan (prevention, quick and strong response to incidents, effective laws and close partnership [234]) to deal with the threat of cybercrime for the government Working with global institutions, other governments, industry partners and Internet Service Provider to analyze and diminish malicious traffic on IT infrastructure. Enhancing the understanding of cybersecurity issues and developing the adoption of good practices in business and communities associations
Developing a vibrant cybersecurity ecosystem	The government cooperates with industry partners, and Institution of Higher Learning (IHLs) to build a cybersecurity workforce with high-tech skills. Developing strong companies and encouraging local start-ups Building a good relationship between academia and industry to develop cybersecurity R&D
Strengthening International partnerships	Building strong international cooperation in cybersecurity, especially ASEAN to identify transnational cybersecurity and cybercrime concerns Building cyber capacity initiatives and exchanging cyber norms and legislation

International cooperation

Beside the cooperation with ASEAN countries, Singapore expanded the cooperation with other organizations such as France, UK, India, Netherland, and the USA [Table 3.5].

Table 3.5: Summarizing the international cooperation between Singapore with the other organizations [235]

Collaboration between Singapore and other organizations	Singapore
ASEAN	Participated in TELMIN, Japan annual engagement, Asia Pacific CERT (APCERT), FIRST (Forum of Incident response and security team), and Meridian process
France – Agence Nationale de la Sécurité des Système d' Information (ANSSI)	Signed MOU (Memorandum of Understanding) Sharing best practices and efforts to develop cybersecurity expertise
United Kingdom	Signed an MOU on cybersecurity cooperation with the cabinet office Joining in cyber research and development
India	Signed an MOU with the department of electronics and information technology of India in 2015 Focusing on five areas: a formal framework for professional dialogue, operational readiness and response, cyber security technology and research in smart technologies, exchanging of best practices, and human resource development exchanges
Netherland	Signed an MOU with the national cybersecurity center (NCSC) in 2016 Sharing cybersecurity best practices and strategies to protect CII and access to training and workshops
USA	Signed an MOU with the National protection and programs Directorate (NPPD) at the Department of Homeland Security (DHS) in 2016 Sharing best practices on CII and cybersecurity trends Building cyber capacity, increasing cybersecurity awareness

In conclusion, Singapore is a developed country and ASEAN's dragon in the international center for trading, finance, and logistics. It is also one of the pioneer nations to realize the advantages of technology to develop itself and the impact of cybersecurity in protecting their critical information infrastructure and cyberspace in order to combat towards cyber threats, cyber-attacks, and cybercrime. Regarding Singapore's cybersecurity strategy, it built up the vision, purposes, and priorities in cybersecurity for Government, businesses, individuals, and the community to ensure the safety of cyberspace. In addition, strong cooperation with international nations or organizations helps Singapore immediately proactive approach and respond to cyber incidents or attacks.

Malaysia

With the purpose to react towards cyber incidents and cyber-attacks, the Malaysian government established the Malaysia Computer Emergency Response Team (MyCERT) under the Malaysian Institute of Microelectronic Systems (MIMOS) Berhad in 1997. Moreover, the Malaysian government set up a cybersecurity center – namely the National ICT security and emergency response center (NISER) on 24 Jan 1998. It was officially adopted on 10 April 2001 under the control of the Ministry of Science, Technology and Innovation (MOSTI). In 2003, Malaysia was a co-founder of the Asia Pacific Computer Emergency Response Team (APCERT) and participated in the Forum of Incident Response Security Team (FIRST). Moreover, the Malaysia government participated in Global business dialogue on electronic commerce (GBDe), Regional Asia Pacific Information Security Standard Forum (RAISS) meetings, and set up a collaboration with the International information systems security certification Consortium (ISC2) in 2001, 2004, and 2005, in respectively [236]. Then, in March 2007, it changed its name to Cybersecurity Malaysia to ensure the safety, safeguard, and development the cyber security in Malaysia [237],[236], [238], [Figure 3.7], [Table 3.6]. Then, Malaysia became the first country in Asia as the chair of the Organization of Islamic cooperation – Computer Emergency Response Team (OIC-CERT) in 2009. Moreover, the Malaysia government set up some technology and innovation centers and programs between 2009 and 2013 such as cybersecurity Malaysia's Malware research center, the security assurance lab, CyberSAFE programs in school, a forensic lab in Malaysia and the Asia Pacific, Malaysia TrustMark for the private sector (MTPS) and Policy and Mechanism for National Cyber Crisis Management for National Security Council in order to create, maintain safer cyberspace, and be recognized as National cyber security reference and specialist center.

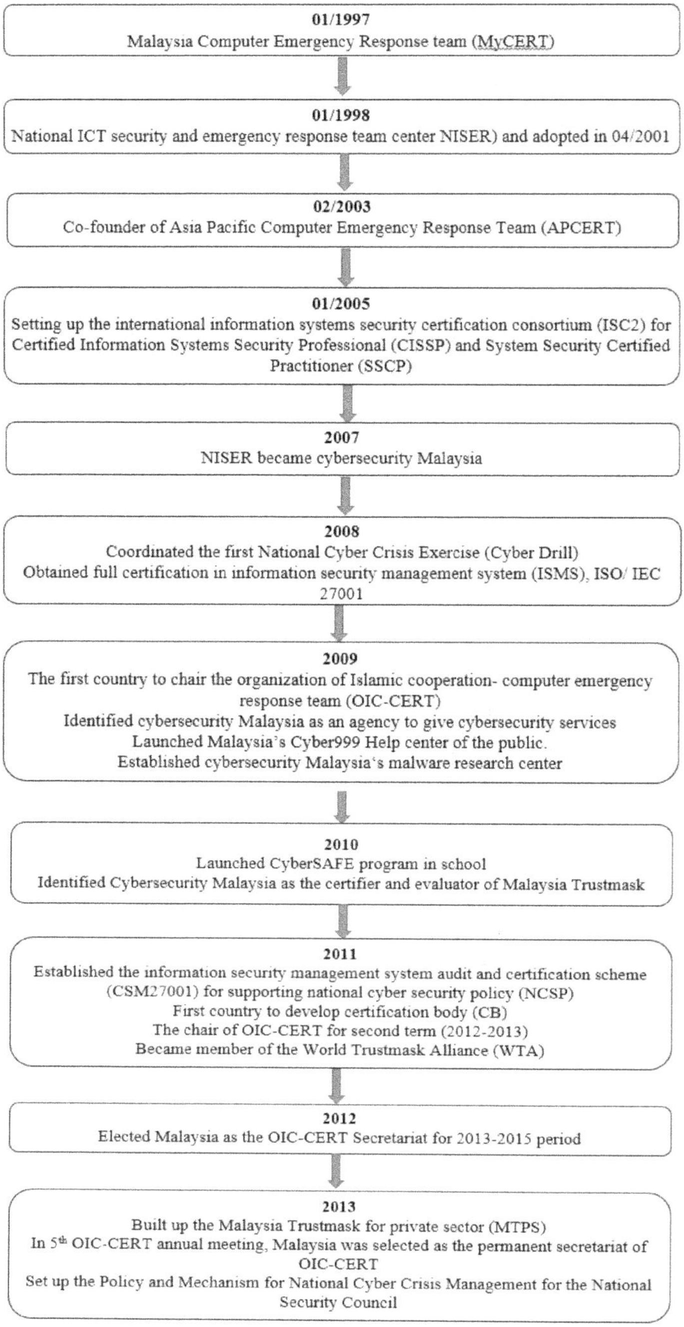

Figure 3.7: Malaysia cybersecurity organizations

Malaysia cybersecurity policy focused on eight important key factors – known as Thrust Frist-Eighth such as effective governance; legislative and regulatory framework; cybersecurity technology framework; culture of security and capacity building; research and development towards self-reliance; compliance and enforcement; cybersecurity emergency readiness; and international cooperation [239][173]. This policy intends to protect the Critical National information infrastructure (CNII) for the most essential sectors of Malaysia following by National defense and security; banking and finance; information and communication; energy; transportation; water; health services; government; emergency services; and food and agriculture.

Table 3.6: Malaysia's cybersecurity services [236], [240]

Cybersecurity's services	Functions
Cyber Emergency Response Team/ Cyber999, CyberDiscovery, CyberDEF	-Addressing cybersecurity issues and cyber threats for Malaysians Internet community such as identifying theft, intrusion, viruses, worms, etc.
Digital forensics (CyberCSI)	- Supporting the regulatory bodies and enforcement agencies for technical assistance with guidance forensic investigation and experts testimonials
Security assurance	-Supporting security evaluation for Malaysian ICT security product manufacturers -Enhancement the status of Malaysian ICT security products and the competitiveness overseas
Security management and best practices (Malaysia Trustmark, MytrustSEAL, CSM-ACE, MyCSC&EDP, MyVAC & MySEF)	-Providing the best practices and standards for organizations and the public to learn, adapt and understand the importance of information security.
Training outreach (CyberSAFE, CyberGURU, e-security bulletin)	-Educating the Internet users on the threats on the Internet
Technical Coordination Centre	- Providing technical coordination and collaboration at the national and international level during a cyber crisis such as a large-scale attack on key information infrastructure
Strategic policy research	- Enhancing research, proposing cybersecurity guideline, and forming an international security framework to reduce the vulnerability of Malaysia's ICT systems and networks - Strengthening Malaysian cybersecurity capability.

In short, the Malaysian government soon recognized the essence of information security; as a result, they created several cybersecurity centers like MyCERT, NISER and joined in some international organizations such as APCERT, FIRST, GBDe, RAISS, ISC2, OIC-CERT, and the like to approach new technology in safeguarding and enhancing cybersecurity. Besides, the Malaysian government increases security awareness for their citizens by boosting education programs in school, developing the technology and innovation centers, research or lab centers, cooperating with private sectors to maintain their cyberspace in order to meet their vision (be realized as National cybersecurity reference and specialist center in global by 2020).

Weak cybersecurity capacity nations

There still exist several countries with weak cybersecurity capacity such as the Philippines, Indonesia, Thailand, Lao PDR, Cambodia, and Vietnam. These countries have suffered many years of war and heavy losses in critical infrastructure, social life, economy, and the like; as a result, they need time to reconstruct their infrastructure, military, system; develop the economy and technology. Hence, they lack experts, technology, and budgets in order to build strong cybersecurity capacity building, strategy, cyber-defense to deal with cyber-threats.

The Philippines

The Philippines is an archipelagic country in Southeast Asia with more than 7000 islands [241]. It has rich natural resources and it is also one of the world's greatest biodiversity-ecosystem nations. Moreover, it is located on the Pacific Ring of Fire and closed to the equator; therefore, it is under the influence of earthquakes and typhoons. In addition, the Philippines is also a honey target for cyber-attacks; indeed, A Frost and Sullivan research by Microsoft said that the potential economic loss because of cybersecurity incidents can reach 3.5 billion USD (approximately 1.1% of Philippines GDP) [242]. As a result, the Philippines Department of Information and Communication Technology (DICT) clarified four main national targets in the National cybersecurity plan in 2005 and it focused on several national priority goals, followed by [243]:

- Creating critical infrastructure trusted and secure
- Securing government information environment
- Safeguarding business
- Enhancing individuals awareness and secure

This national plan clarified four main parts to prevent cyber-attacks for the Philippines; for example, define cyberspace, challenges in critical cyber infrastructure protection, international and domestic cybersecurity regime, and the national plan for critical cyber infrastructure security [244]. The Philippines government expected this plan might standardize the necessary capabilities in the government and private sector, as a result, mitigate and respond to the cyber threats or cyber-attacks against the national critical infrastructure. Since the appearance of the Department of Information and Communication Technology, it can help the Philippines government standardize the adoption and implementation of Information security governance and risk management approaches via the establishment of the National Computer Emergency Response Team (NCERT). Moreover, the National Cyber Security Plan 2022 (NCSP) of the Philippines was established in May 2017 with four primary purposes as follows: ensuring the continuous operation of the Philippines critical info-structure, public and military networks; applying cyber resiliency measures to improve the ability to mitigate threats before, during and after attacks; making effective coordination with law enforcement agencies; and educating cybersecurity society [245]. In order to protect the National critical info-structure, government in public and military, small and large businesses, and everyone using the Internet. Furthermore, The Philippines government created some special organizations in order to fight against cybercrime and cyber-attacks such as [245]:

- Department of Information and Communications Technology
- Cybercrime Investigation and Coordination Center (CICC)
- National Computer Emergency Response Team (NCERT)
- Department of Justice – Office of Cybercrime
- National Bureau of Investigation – Cybercrime Division
- Philippine National Police – Anti-Cybercrime Group

Challenges in Philippines cybersecurity strategies [242]

In the research of A Frost and Sullivan [242], it pointed out that there were three main challenges in the Philippines organizations' cybersecurity strategies that can lead to data corruption, data breaches such as

- *Security an afterthought*: only 44% of firms recognized cybersecurity before the digital project begins, and 56% thought about it after the project starts or do not consider it at all.
- *Creating a complex environment:* 17% of respondents with more than 50 cybersecurity resolutions to recover damage from the attacks during an hour but 38% with less than 10 solutions one.
- *Lack of cybersecurity strategy*: 46% of respondents think of cybersecurity strategy as a means to protect the company against cyber attacks rather than a strategic business promoter.

Simultaneously, the lack of IT security experts is also one of the weaknesses of the Philippines. For example, there are only 84 experts in Certified Information Systems Security Professional (CISSP) but nearly half of them

(40) are working abroad [246]. Likewise, in comparison with other countries 'certified experts, the Philippines still has a big gap; for instance, with Singapore more than 1000 experts, Indonesia with 107 specialists, Thailand with 189 professionals, and Malaysia with 275 ones [247].

International cooperation

The Philippines has a strong relationship in security cooperation with the USA in counterterrorism or maritime security [248]. Indeed, between 2002 and 2015, the USA deployed more than a hundred special operations and practices to the Southern Philippines for fighting against terrorism goals. Besides, the Philippines was also a member of the Association of Southeast Asian Nations (ASEAN) from 1967 [249]. The main cooperation is in economic, social, cultural, technical, educational, and other fields. Furthermore, the Philippines is also a close partner with Japan in maritime security cooperation because both of them are maritime nations and another reason is that Japan wants to reduce the influence of the Chinese in the geopolitically strategic Southeast Asian nations [250]. Likewise, the Philippines CERT also cooperated with many international organizations in the same region such as Cybersecurity Malaysia, MyCERT, Microsoft, and so on to increase their safety for national security and cyberspace.

Indonesia

Indonesia is a country which made up of thousands of volcanic islands. It is also a nation gathered of ethnic groups with many different languages. Moreover, it is known as the second most targeted nation for cyber-attacks (approximately 50000 cyber-attacks every day) [251]. However, Indonesia is at the beginning stages of developing a national cybersecurity strategy. The legal framework is still weak and there are no precise security laws, policies, security practices, and specific cybersecurity plans [252]. Nonetheless, the Indonesian government created some cybersecurity organizations to mitigate cybersecurity issues [Table 3.7].

Table 3.7: Indonesia cybersecurity organizations [253], [254], [255]

Cybersecurity organizations deal with cybersecurity issues	Functions
Indonesia computer emergency response team (ID-CERT) in 1998	Using for the public sector and works based on complaints
Indonesia security incident response team on internet infrastructure (ID-SIRTII) in 2007	Secure the use of telecommunication networks based on internet protocol
Academic CSIRT (Acad-CSIRT) in 2010	Focusing on the development of security in Indonesia (for State and private universities)
The Directorate of information security in 2011	Formulating and implementing policies, technical standards
Government computer security incident response team (GovCSIRT) in 2012	Cooperating with ID-CERT and ID-SIRTII to monitor, evaluate, incident response, and develop security capability of government stakeholders
Indonesian National Police (POLRI)	Cybercrime unit – responsible for law enforcement and policing duties
Ministry of laws and human rights	Responsible for information technology and electronic transactions, telecommunications, and intellectual property
National cyber information defense and security	Strengthening cyber warfare and cyber defense capabilities
Desk at the coordinating ministry for political, legal, and security affairs	-Planning and policy coordination -Synchronizing policies in the aspects of politics, law, and security
Badan Cyber Nasional - BCN (National cybersecurity agency)	- Managing the State cryptography agency, the State intelligence agency

Besides, Indonesia also cooperated with some international organizations in countering the cyber-attacks and cybercrime, follows by:

International cooperation

- A member of ASEAN Network Security Action Council and International
- International Telecommunication Unit (ITU).
- The steering committee of Asia Pacific Computer Emergency Response and Security (APCERT).
- Having bilateral cooperation with Japan, the United Kingdom, and other countries
- Cooperation between Bandung Institute of Technology (ITB) and the Korean International Cooperation Agency (KOICA) to support cybersecurity education training, and research [256], [253]

Obstacles for Indonesia national cybersecurity [257], [258]

- Lack of awareness in information security
- Cyberlaw and policy aren't complete
- Governance and organization of national cybersecurity are weak
- A limitation export market
- Lack of human resources both quantity and quality in information security
- Coordination and cooperation between agencies
- ICT critical infrastructure protection mechanism and standard not exist
- Application, data, and infrastructure of information security not integrated
- Software piracy, weak support for R&D

Thailand

In 1992, Thailand's government set up the National Information Technology Committee (NITC) with the main task of converting policies into actions and practices to develop the Thai economy and society. In order to facilitate the implementation of policy, government agencies such as the National Electronics and Computer Technology Center (NECTEC) and Software Park were found [259]. It was a government organization by the National Science and Technology Development Agency (NSTDA) and it was also the secretariat of NITC. Then, in 1996, the Thailand government established the first National Information Technology Policy- called IT2010. This ICT Policy framework (IT2010 policy) is considered as a long-term policy at the macro level with three key areas for IT development such as investing for national information infrastructure, investing in human resources development, and good governance (enhancing the government services) [260]. Moreover, it also emphasized five main strategic fields in the development and application of ICT, namely e-government; e-industry; e-commerce; e-education; and e-society in order to improve the economy and quality of Thai citizen's life [260], [259]. This policy clarified three major purposes, as follows:

- Improving Thailand's ranking in the Technology Achievement Index (TAI) from "dynamic adopters" group to "potential leaders" countries.
- Increasing Thai skilled workers to 30 percent of the workforce by 2020
- Enhancing the Thai industry towards the knowledge-based industry to reach 50 percent of GDP.

After establishing IT2010, Thailand's government started to focus on building some information security Acts for business such as the electronics transaction Act B.E.2544, computer crime Act B.E.2550, and electronics transaction Act (2nd Amendment) B.E.2551 to protect the business transactions in 2002, 2007,

2008, respectively. Afterward, the Thailand national IT committee began to build the IT2020 policy and drew electronics transactions and a digital masterplan [Figure 3.8].

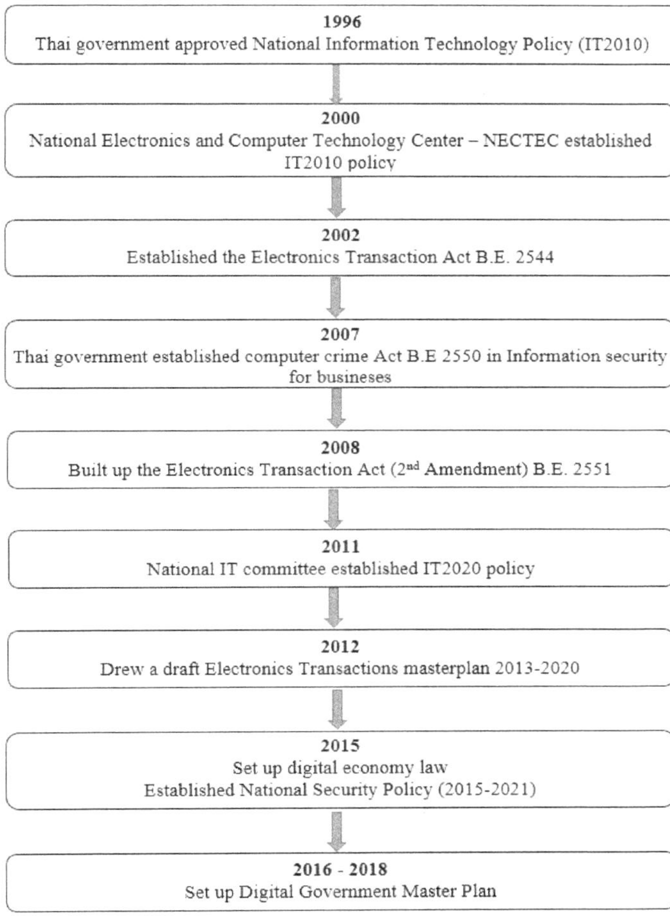

Figure 3.8: Thailand cybersecurity development

With the IT2020 policy, the Thailand government intended to develop their country as a smart development nation based on knowledge, wisdom in economy and society. They emphasized that ICT is a key to lead Thai people to reach knowledge, wisdom and develop society towards equality and a sustainable economy in the same region [261]. Beside the development of IT2010 and IT2020 policy, NECTEC established the Computer Emergency Response Team (ThaiCERT) in 2000. ThaiCERT is also the Computer Security Incident Response Team (CSIRT) for dealing with computer incident reports in the Thailand Internet community. ThaiCERT has been the first and only non-profit CSIRT in Thailand [262]. In February 2011, ThaiCERT operations were transferred to a new administrative team in a new public organization, namely Electronic Transactions Development Agency (ETDA) under the supervision of the Ministry of Information and Communication Technology. Furthermore, ThaiCERT cooperates with the Thai government sector, organizations, universities, ISPs, and other relevant entities to manage computer security incidents in Thailand. In addition, ThaiCERT is also a member of the Forum of Incident Response and Security Teams (FIRST) and the Asia Pacific Computer Emergency Response Team (APCERT) and it cooperates with global and local CSIRTs in responding to computer security incidents. Currently, in 2018, the Prime Minister of Thailand organized the meeting on cybersecurity to develop a cybersecurity agency with several purposes of making Thailand among the top 20 countries in the world of cybersecurity readiness, following by [262]:

- Creating the national policies framework that safeguards, limits, and reduces cybersecurity threats.
- Developing the Critical Information Infrastructure (CII), creating the guidance and Standard Operations Procedure (SOP) in some cybersecurity emergency cases.
- Enhancing the cybersecurity personnel
- Building the Cybersecurity Agency (CSA) responsible for countering cybersecurity problems and protecting the country's national cybersecurity.

Lao People Democratic Republic (PDR)

Lao PDR is located as a country in the center of Southeast Asia. It has the same borders with five surrounding countries; for instance, China, Cambodia, Vietnam, Myanmar, and Thailand. Laos is one of the poorest countries in Asia with 27 percent of citizens who are living less than one dollar per day [263]. Laos's government recognized that ICT can improve the development of the country; however, the Lao People Democratic Republic (Lao PDR) had experienced similar kinds of cyberattacks like the other countries in the same region and other parts of the world. Therefore, in 2009, the national ICT policy was established. Before 2012, Lao PDR was the only one in ASEAN countries which didn't have a National Computer Incident Response Team (Nation CIRT). Nevertheless, regarding the increasing number of cyberattacks and the quick boosting of ICT, in February 2012, the Lao Computer Emergency Response Team (LaoCERT) was established and recognized as one division under the Lao National Internet Center (LANIC) [264], [265]. Moreover, based on the recommendation of the International telecommunication Union – the International Multilateral Partnership Against Cyber Threats (ITU-

IMPACT) [266], in June of 2016, LaoCERT was divided from Lao National Intern Center to become a National CERT of Lao PDR and under the monitoring of Ministry of Post and Telecommunications. At present, LaoCERT is a member of Asia Pacific CERT (APCERT) with 4 divisions such as administration and cooperation, research and development, technical, and information monitoring [267]. Furthermore, LaoCERT also enhances the collaboration with some regional organizations; for example, took part in ASEAN - Japan activities in 2012, signed the MoU with ThaiCERT in 2013, IDSIRT in 2015, VNCERT and CNCERT/CC in 2017, Cambodia Computer Emergency Response Team (CamCERT), Japan Computer Emergency Response Team (JPCERT), and FIRST to improve ICT environment secure and safe [267]. Last but not least, Laos is the first country in the ASEAN group to sign the MOU on 19, June 2018 about the usage of Blockchain technology with Lina Network Corporation in order to do research and develop "Digital Identity" for Laos's government. With this technology, Laos's government enhances in managing the citizen's data flow absolutely, ensuring privacy as well as identity management and authentication information with simple applications [268].

Legislation and laws

Lao government created the policy to enhance the security of the ICT sector as a critical tool for social and economic development with laws, regulations, decrees, and related legislations, following by [269], [270],[271], [272]:

- National ICT policy (2009)
- Telecommunication law (21/12/2011)
- E-transaction law (7/12/2012)
- Criminal law (11/12/2012)
- Draft of National Broadband plan (2012-2020)
- Draft e-government master plan (2013-2020)
- Decree on online information management (2014)
- Cybercrime law (15/7/2015)
- Draft of National ICT policy (2015-2025)
- ICT law (2016)
- Drafting Data protection law (2017)
- Ministry Post Telecommunication (MPT) vision 2030, strategy 2025, and development plan 2020

ICT policies

Lao PDR clarified nine major areas in ICT policies with a long term consideration such as Infrastructure and Access; Enterprise and Industry; Research and Development; Applications; Human Resource Development; Legal Framework; Awareness; Poverty Alleviation; and Standardization and Localization [258],[273], [Table 3.8].

Table 3.8. Lao's ICT policies

ICT policies areas	Functions
Infrastructure and Access	- Focusing on expanding the existing telecommunications infrastructure - Linking rural and remote areas - Providing telecommunication services to underserved areas - Reducing import tax for ICT equipment
Enterprise and Industry	- Encouraging enterprise development in the ICT sector - Supporting national and foreign investors to compete and cooperate in investment in ICT fields - Promoting local ICT enterprise development by reducing tax, import ICT equipment
Research and Development	- Developing national research and development centers in ICT - Supporting cooperation with private sector ICT companies - Encouraging the development of the National ICT Association (NICTA)
Applications	- Enhancing in providing some services and management like e-government, e-tourism, and banking
Human Resource Development	- Promoting and supporting ICT learning to ensure the necessary capacities to meet national goals - Creating the telecentre programs to enable ICT learning in rural and remote areas - Building up the exchange of technical knowledge and expertise
Legal Framework	- Developing a comprehensive of cyber-laws to manage information networks - Encompassing e-commerce/ e-business, cybercrimes, consumer protection, and intellectual property rights
Awareness	- Implementing a public awareness program to ensure citizen's awareness of ICT importance - Encourage the private sector and the international community to support the public awareness program
Poverty Alleviation	- Safeguarding the growth with equity (gender, ethnicity, location, and returnee status) - Facilitating the application of ICT on social networks (civil society, academia, the general public, government, and private sector) -Focusing on environment, health, gender, and youth

ICT policies areas	Functions
Standardization and localization	- Developing software, hardware, protocol standards, equipment services to ensure interoperability and harmonization with international, regional, and sub-regional standards - Establishing the network on ICT (national and international experts, academia, government, and the private sector) - Adopting Unicode standard for Lao script, and improving digital interchange in the Lao language

Although there are some limited in development of IT skills; human resources; infrastructure development; capacity building; and finance, Laos' ICT has fully supported by the government in order to fight against cybercrime.

Cambodia

Cambodia is a slowly developed country in Southeast Asia with the lowest Internet connection in the same region. Based on the researchers, policy makers, and international stakeholders, they recognized that ICT could help this small country to narrow the gap with the global digital environments as well as other countries in ASEAN. Internet first started in Cambodia with the commercial service in 1997. After four years later, the number of Internet users in Cambodia was still 8000, approximately 0.07% of its population [274], [275]. National ICT Development Authority (NiDA) was responsible for ICT development in Cambodia and it has been linked to the Ministry of Posts and Telecommunications (MPTC)'s structure [275]. Moreover, Cambodia Computer Emergency Response Team (CamCERT) was established in 2007. It is an office under the Information and Communications Technology (ICT) Security Department and Ministry of Posts and Telecommunications (MPTC).

Cambodia's ICT Masterplan for 2020 purposes to create an "ICTobia" which provides the country's development toward intelligence [276]. This Masterplan focuses on five prior goals such as "empowering people, ensuring connectivity, enhancing capabilities and enriching e-services" [277], [Table 3.9].

Table 3.9: Cambodia's ICT Masterplan by 2020 [277]

Objectives	Aims
Empowering people	- Becoming a top-tier country in Southeast Asia in ICT human resource development - Gaining 70% of Cambodian citizens access the Internet by 2020
Ensuring connectivity	- Enhancing services accessibility of telecom and broadcasting for people - Widening ICT structure via government assistance - Enabling private investment and setting the standard for diverse ICT - Building national ICT infrastructure, legal framework, and cybersecurity
Enhancing capabilities	- Standardizing ICT - Cooperating national ICT ecosystem to global ecosystem - ncreasing the number of participants - Raising up ICT technology capacity via R&D to reinforce national competitiveness
Enriching e-services	- Evolving an e-government framework, increasing cybersecurity, e-education, e-commerce, e-public services, and e-tourism

Regarding this ICT masterplan 2020, the Cambodia government declared five projects which may enrich e-services in a short term (developing technical framework for Cambodia government, enhancing for establishing ICT security) and long-term plans (promoting e-commerce, establishing tourism network, and developing educational program) [277]. Moreover, the Cambodian government applied the Law on telecommunication in 2015 and began its ICT development policy in 2016. Likewise, they also drafted several legislations like e-commerce and cybercrime law [275].

A case of Viet Nam
E-government and E-commerce

E-government

In 2010, it was a remarkable year in the development of e-government in Vietnam. Regarding the implementation of Decision 43/2008/GD-TTG and 48/2009/QD-TTG of ICT application in state agencies period 2011-2015, the government invested approximately 1700 billion Vietnamese currency [278]. Vietnamese e-government mainly paid attention to four main target clients such as individuals, enterprises, governmental officials, and governmental agencies [279]. It can help Vietnamese officials to diminish time and expense; reduce stagnation, bureaucracy, and extortion; operate 24/7; satisfy the demand of social needs; increase transparency and decrease paper, and so on [280]. During 26 years, the Vietnam government implemented 5 big projects, two of them was supported by the French government (in 1991-1993 and 1994-1996); one was provided by the State budget (1996-1998), another one was under the Prime Minister's Decision in 1997 and the last one was considered as the milestone for e-government in Viet Nam from 2001 to 2007. Although all achievements were not as successful as expected [278], Vietnam's position rank has increased every year regarding the global rank of e-government readiness [281]. However, in 2008 -2010, the Vietnam government established Decree 64 to enhance the government capability's management, offer some e-services, and develop IT human resources. Then, in the period from 2011 to 2015, the e-government system was quite complete with all basic public e-services such as online register, license, payment, and so on. By the year 2020, Vietnam's e-government will be expected to a ubiquitous government (U-Gov) system anywhere, anytime, and on any devices [280].

E-commerce

Vietnam has had several typical systems such as Vietnam cyber mall, real estate exchange, e-business, blue sky, bookstore, electronics, and mechanical appliances supermarket and so on [282]. Vietnam's e-commerce is quite new [283]. It lacks e-commerce law which is one of the barriers for foreign companies in trading with Vietnamese firms. Therefore, at the 4th ASEAN summit meeting in Singapore (Nov 22nd to 25th, 2000), Vietnam signed the e-ASEAN framework agreement to facilitate e-trading in ASEAN [124]. Moreover, the Vietnamese Political Bureau promulgated a Politburo's Directive No.CT58BCT on Oct 17th, 2000, followed by the government's decision No 81/2001/QDTTG to develop information technologies in the cause of industrialization and modernization [124]. With the objectives in the year 2020, Vietnam's ICT will reach the advanced level in the region to make the economic branch increase at the high growth rate in order to contribute to the Gross Domestic Product (GDP) growth. In order to achieve these objectives, the Vietnamese government implemented several programs following:

- Building and improving the telecommunications and Internet infrastructures
- Development of the IT manpower resource
- Establishing and enhancing the software and hardware industry

Network security incidents

Cyber-attacks are becoming more sophisticated and they are the greatest threats for every organization in the world. It causes not only financial losses but also operational interruption [284]. Network security in Vietnam has many vulnerabilities holes in the airport system, banks, websites and the security status is now at a high warning level.

Indeed, in 2016, there was a huge attack on Vietnamese airplane websites, especially at several international airports like Tan Son Nhat, Noi Bai, Da Nang, and Phu Quoc. It was attacked by a hacker group (referred to 1937CN) from China and this attack made data leakage of more than 400,000 member accounts [285], [286], [287]. Moreover, it interrupted the check-in process at the international airports and it made many airplanes need to delay for a few hours. According to the Vietnam Computer Emergency Response Team (VNCERT) report in 2017, Vietnam had 13,382 cyber-attacks including malware, phishing attacks, and deface attacks [288]. One year later, VNCERT also reported that Vietnam was under attack by 6,500 cyber-attacks during eight months of 2018. Almost all attacks are the Distributed Denial of Service attack (DDoS) to collect data from government websites and offices [289]. In order to prevent and respond or mitigate cybersecurity incidents, the Vietnamese government clarified the responsibilities of each organization in operational entities to establish the cyber laws, Decrees, or the Acts to deal with them.

Operational entities

The Vietnamese government has several cybersecurity organizations responsible for cybercrime, cyberwar, and cyber-attacks as the Ministry of Public Security, the Ministry of Information And Communications, and the Ministry of Defense [Figure 3.9], [290]. Firstly, the Ministry of Public Security has three main entities: Department of Network Security (namely "A68"), Department of Information Security and Communication (namely "A87"), and Police Department of Prevention and Fight against High-Tech Crime (namely "C50"). They are responsible for the management, control of information system security and cybersecurity, and encounter online fraud, financial crime.

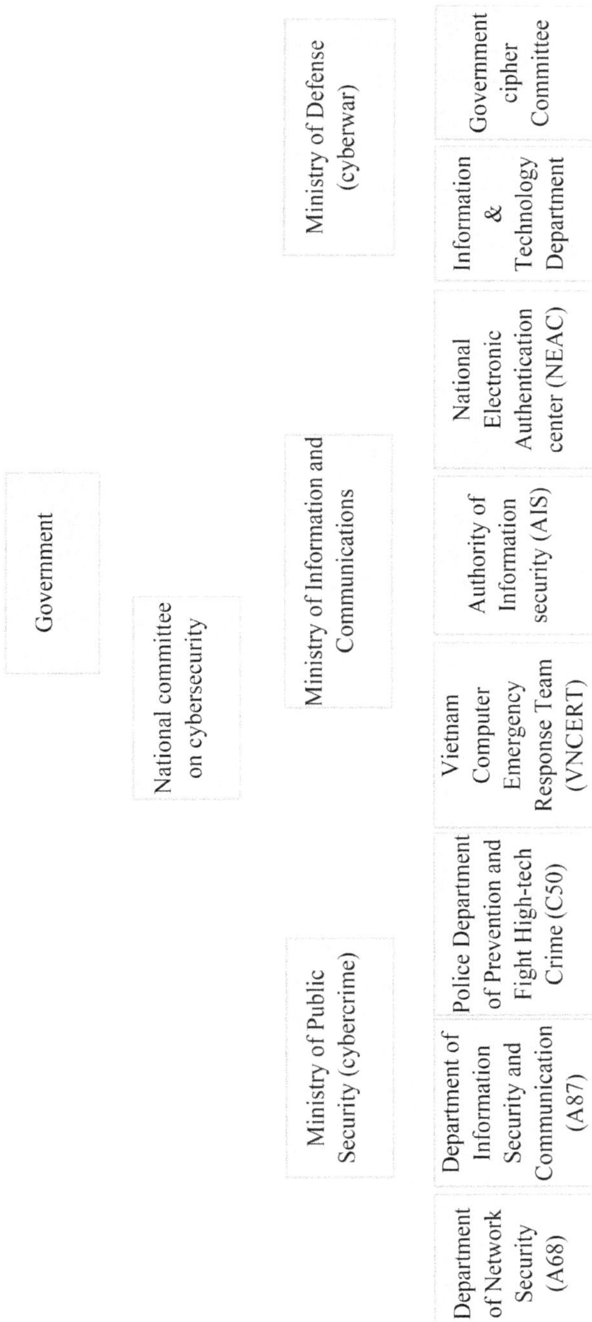

Figure 3.9: Vietnamese cybersecurity organization [290]

Secondly, the Ministry of Defense (MOD) has two main departments: The information and Technology Department and Government Cipher Committee. They are under the control of the Joint General Staff of the People's Army of Vietnam and the Minister of Defense. Moreover, they are dealing with managing encryption communication and networks, strategy, policies, and legal documents; as well as applying encryption solutions, products, and improving development and research. Remarkably, the Ministry of Information and Communications manages three main parts VNCERT, Authority of Information Security (AIS), and National Electronic Authentication center (NEAC) to ensure cybersecurity for the nation and civilians. Besides, the private sectors and Vietnamese companies in ICT also play an essential role in improving the safety of critical infrastructure systems and cyber resilience capacity; for example, the Vietnam Information Security Association (VNISA), the Vietnam Software and IT Services Association (VSISA), the Vietnam Internet Association (VIA) and the Vietnam E-commerce Association (VEA), the Vietnam Association for Information Processing (VAIP). They are the key factors to encourage the R&D and offer cybersecurity solutions, products, or services not only for the government but also for citizens.

VNCERT

Vietnamese computer emergency response team (VNCERT) was established in 2005. It is an official organization of the Ministry of information and communication. This organization manages computer incidents, alerts network security issues, builds network safety standards, and promotes building CERT systems in other organizations, companies, or businesses [Figure 3.10]. Moreover, it also cooperates with other international CERTs. For instance, in November 2018, VNCERT collaborated with PricewaterhouseCoopers (PwC) in Vietnam to organize a cybersecurity drill namely "Enhancing analytical, investigate and response skills to deal with cybersecurity incidents" in three big cities Ho Chi Minh, Hanoi, and Da Nang [291]. This cyber drill aimed to investigate the network vulnerabilities, attack tactics; and exchange the knowledge, experiences, and skills between the experts towards Advanced Persistent Attack (APT). As a member of the Cybersecurity Incident Response Team (CSIRT), VNCERT often organizes international cyber drills and exercises to promote the information security capabilities for technical staff in all organizations. In fact, it holds the cyber drill for ASEAN CERT's Incident Drill (ACID), Asia Pacific CERT's (APCERT) drill, and ASEAN-JAPAN drill every year.

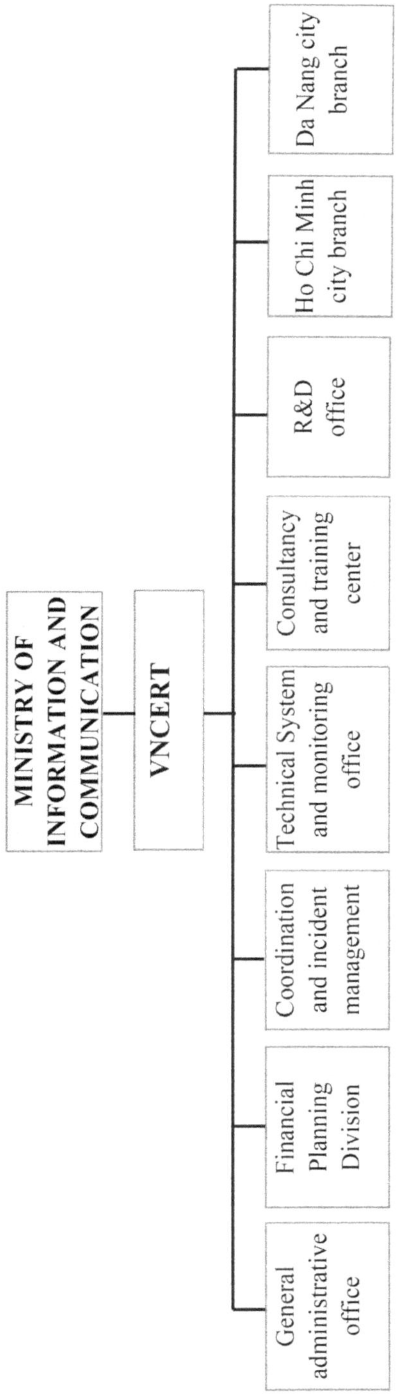

Figure 3.10: VNCERT structure [292]

Legal foundations

Vietnam has a marvelous process in Information communication technology over the past decades in comparison to the other nations in the same region. However, the Vietnamese government quite less paid attention to cybersecurity at the beginning state. In fact, the Vietnamese government just has several cyber-laws to protect information during a transaction on information exchanges, e-commerce, and foreign trading between individuals, organizations, and government officials. Firstly, they established the law on e-transaction, the law in information technology, and law in telecommunication in 2005 and 2006, respectively [293], [294], [295]. Vietnam has a marvelous process in Information communication technology over the past decades in comparison to the other nations in the same region. However, the Vietnamese government quite less paid attention to cybersecurity at the beginning state.

Firstly, they established the law on e-transaction, the law in information technology, and law in telecommunication in 2005 and 2006, respectively [293], [294], [295]. Continuously, the Vietnamese government offered several Decrees like in 2007 on the application of information technology in State agencies' operation ("Decree No. 64") and 2008 for anti-spam to enhance the country's cybersecurity capability ("Decree No.90") [290]. Then, until 2016, they had special laws on cybersecurity such as government Decree No.85 in July 2016 on the protection of information system [296] and government Decree No. 108 on conditions for provisions of cybersecurity products and services [297]. Regarding these decrees above, they gave the guidelines to ensure the security of information and products or services during their operations. Furthermore, there was the Law on cybersecurity found by National Assembly in 2015, and adopted in 2016 [298] to enhance network information security activities; declare the responsibilities of agencies, businesses, organizations and individuals in protecting network information security; and define technical standards or terms in the cyberspace and state management. Recently, the latest Cybersecurity Law (the CSL 2018) has just enacted in 2018 and it was adopted in January 2019 [299]. This law provided the protection of cybersecurity for all agencies, organizations, and individuals. It involves all elements of Vietnam IT infrastructure such as "telecommunication, Internet, computer systems, databases, information processing, storage and controlling system and related activities of Internet Service Providers in cyberspace" [299]. It also aimed to protect both critical information systems and non-critical information systems.

International cooperation

Vietnamese government step by step improves cybersecurity laws to enhance national cybersecurity for national information infrastructure via the Decrees, the Laws, and the Acts. Besides, they also recognize that the Internet has no border and cyber-threats are the global challenges not just only for a nation. Therefore, developing cooperation with other organizations in the same region or outside is an essential thing. For example, in 2007, the Vietnam government and the Czech Republic government have signed a Memorandum of Understanding (MOU) on cybersecurity cooperation. This bilateral cooperation is a key factor to help Vietnam to enhance national development and modernization, strengthen the nation's competitiveness and international integration, and guarantee the sustainable development of information security [300]. Similarly, in 2014, VNISA – a non-profit organization signed the MOU with Microsoft to enhance information security and privacy in Vietnam. With this cooperation, it strengthens the practical training exercises to increase the capabilities for handling information security issues, boosting the information security services market, and approach new information about security incidents, threats, or attacks. Likewise, in order to tie the relationship between Vietnam and India, they signed the MOU between VNCERT and Indian Computer Emergency Response Team (CERT-IN) in the field of cybersecurity for information exchanging about knowledge and experiences in preventing, detecting, and resolution of cybersecurity incidents in 2016 [301]. Furthermore, PwC in Vietnam and VNCERT signed MOU and started a strategic partnership from 2018 to 2020 to develop a national cybersecurity emergency response network and promote training activities against cyber-attacks [302]. Similarly, Vietnam's RMIT University and the Netherlands Organization for Applied Scientific Research (TNO) also signed the MOU to strengthen cyber expertise in IoT, blockchain, and dark web to increase cybersecurity awareness and best practices in students and teaching staffs towards global cyber-threats [303]. Recently, the Vietnam government also cooperates with the global cybersecurity company – Kaspersky Lab to help the Vietnamese government increase cybersecurity capacity; indeed, The National Cyber Security Center (NCSC) has just signed a Memorandum of Understanding (MoU) with Kaspersky Lab in 2019 [304]. Regarding this agreement, they can share knowledge, technical capabilities, and best practices to ensure the information security of individuals, businesses, and government organizations. Moreover, this cooperation also helps the Vietnamese government develop cybersecurity capacity, institutions, and infrastructure to safeguard public safety and security [305].

Education

Currently, cyber-threats are very complicated towards all countries in general and Vietnam in specific. As a result, the Vietnamese government established several Decrees and programs to promote cybersecurity awareness and human resources for the nation. In fact, they gave Decree No. 99/QĐ-TTG and 153/QĐ-TTg to develop cybersecurity human resources; attract experts or students, individuals in government offices; and increase the number of students studying abroad in ICT from the period 2014 to 2020 [306], [307]. Moreover, the Vietnam Information Security Association (VNISA) also organized annual national contests, conferences for students of all universities and colleges in order to introduce artificial intelligence to safeguard cybersecurity and information security in ICT, IoT, and protect the critical databases or infrastructure [308]. On the private sector side, Bach Khoa Antivirus (BKAV) – a company that was found in 1995 in Vietnam, referred as a leading company in network security, software, and producing smartphone or smart home devices. It also released the first cybersecurity training program online for all people, businesses in 2015 with the purposes to develop the force of cybersecurity in Vietnam, and upgrading comprehensive knowledge on Internet security, cybersecurity as well as attacks and prevention from them [309].

In summary, Vietnam is a developing country that quickly approaches in ICTs and innovative technologies but it is a newbie in cybersecurity protection. With a series of cyber-attacks on government, companies, agencies, and airport websites; they made huge damage to data loss, data leakage, and finance. Hence, the Vietnamese government paid attention to making cyber laws, legal documents, and legal infrastructure to ensure the safety of critical infrastructure protection. Regarding the connection between government organizations and private sectors (VNISA, VSISA, VIA, VEA, and VAIP), it helps to strengthen the safety of critical infrastructure systems and cyber resilience capacity, develop research and training, and promote cybersecurity solutions, products, or services. Besides, the Vietnamese government also considered the important role of international cooperation as a key factor to boost cybersecurity development to a new level in the same region.

The differences in cybersecurity capacity between ASIA and ASEAN nations

Results of cybersecurity capacity in ASIA countries

	Cybercriminal legislation	Cybersecurity legislation	Cybersecurity training	**LEGAL MEASURE**	National CERT/CIRT/CSIRT	Government	Sectoral CERT/CIRT/CSIRT	Standards for organizations	Standards for professionals	Child online protection	**TECHNICAL MEASURES**	Strategy	Responsible agency	Cybersecurity metrics	**ORGANIZATIONAL**
China	◎	○	○	◎	○	○	○	◎	◎	○	○	●	○	●	◎
Hong Kong	-	-	-	-	-	-	-	-	-	-	-	-	-	-	-
Japan	○	○	○	○	○	○	○	◎	○	○	○	○	○	●	○
South Korea	◎	○	○	○	○	○	●	○	○	○	○	●	○	○	○
North Korea	-	-	-	-	-	-	-	-	-	-	-	-	-	-	-

	Standardization bodies	Cybersecurity good practices	R&D programs	Public awareness campaigns	Professional training courses	Education programs	Incentive mechanisms	Home-grown industry	**CAPACITY BUILDING**	Bilateral agreements	Multilateral agreements	International participation	Public-private partnership	Interagency partnerships	**COOPERATION**	**GCI**
China	○	○	○	○	◎	◎	○	○	○	◎	●	○	●	○	◎	◎
Hong Kong	-	-	-	-	-	-	-	-	-	-	-	-	-	-	-	-
Japan	○	○	◎	○	○	○	●	○	○	◎	○	○	●	○	○	○
South Korea	○	○	○	○	○	○	◎	○	○	●	●	○	○	○	◎	○
North Korea	-	-	-	-	-	-	-	-	-	-	-	-	-	-	-	-

Notes: ○ : the highest, - : no information, ◎ : low, ● : the lowest

Figure 3.11: Global cybersecurity index 2017 of ASIA and PACIFIC region scorecard [112]

Regarding [Figure 3.11], it can be seen that Asian nations like China, Japan, and South Korea have well-structured organizations in cybersecurity. For instance, they established legal frameworks to prevent cybercrime and practice cybersecurity training. The most important thing is that they have stronger data protection regulations than European countries such as China, or Hong Kong. In fact, their data protection regulations restrict the data for the third party outside the border. Furthermore, these countries also had strong capacity building such as best practices, R&D programs, public training courses, and the like to enhance the cybersecurity inside. Likewise, they also built several cybersecurity teams like CSIRT, CERT, Gov-CERT, and CIRT to handle cyber incidents for organizations and individuals. However, their public-private partnership and bilateral agreements in these countries with international cooperation were quite low. The main goal of these countries is that they not only want to protect their national security but also they want to promote their position in cybersecurity aspect with the other countries in the same region; therefore, they focus on building capacity, sharing knowledge, creating cyber-laws, data protection regulations or legal legislation, and so on to mitigate cyber-threats and reduce the damage of cyber-attacks.

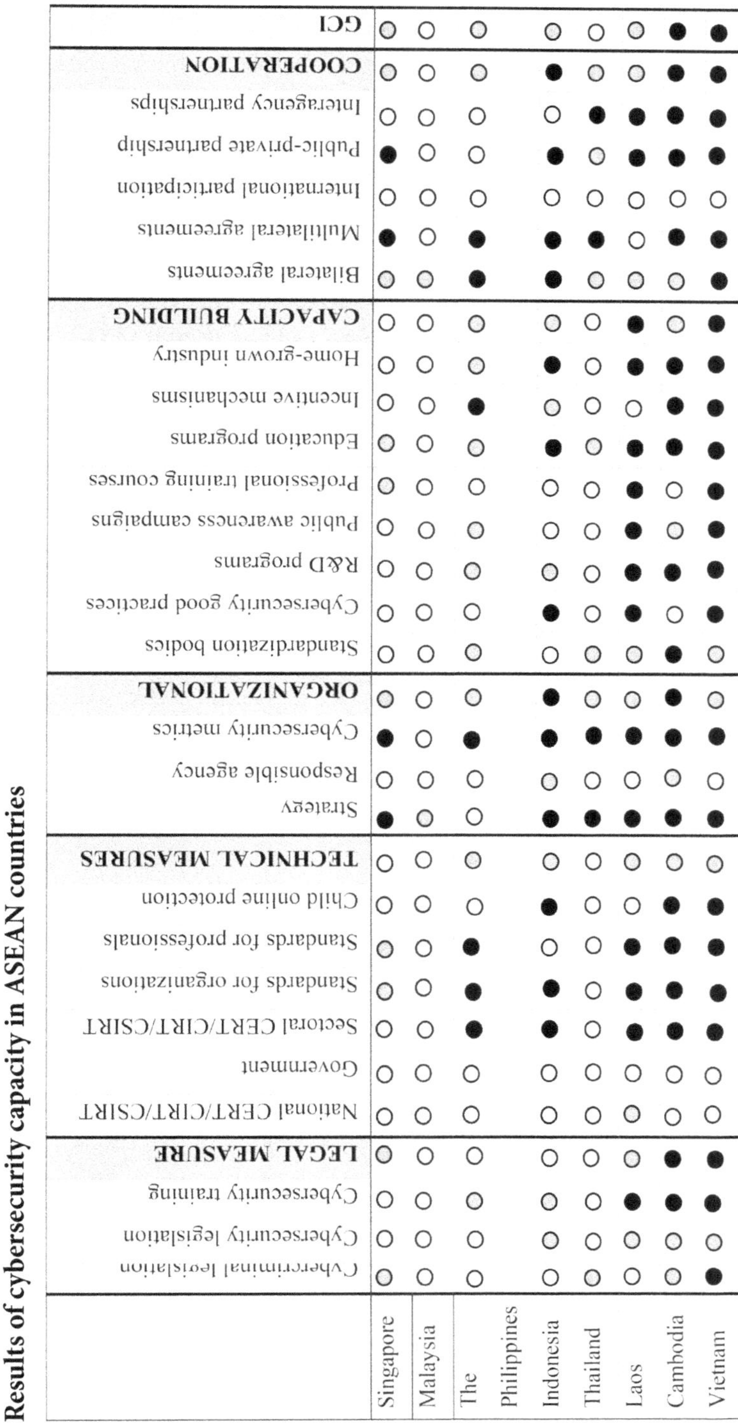

Notes: ○ : the highest, ◎ : low, ● : the lowest

Figure 3.12: Global cybersecurity index 2017 of ASEAN scorecard [112]

As can be seen in [Figure 3.12], Singapore and Malaysia are the strongest countries in ASEAN in capacity building, legal measure, technical measure and cooperation in the same region. In addition, their cybersecurity capacity is nearly equivalent to Japan, China, and South Korea. In another hand, Cambodia, Laos, and Vietnam are the weakest nations in every aspect in cybersecurity capacity building. These nations suffered heavy consequences from the war in the past for many years; therefore, it influenced their economic development, social life, especially in technology development. This leads these nations to take a lot of time to reconstruct the infrastructure system, develop the economy, military, capacity building, and technology. As a result, their cybersecurity capacity building is the lowest in the same region. Besides, the lack of expert, technology, and budget are also important problems for the less digitally developed nations to build strong cybersecurity strategy and capacity building in cybersecurity or cyber-defense.

Table 3.10: Global cybersecurity rank in 2017 of Visegrád, ASIA and ASEAN countries

Visegrád countries	**Score**	**Global Rank**
Poland	0.622	34
The Czech Republic	0.609	35
Hungary	0.534	51
Slovakia	0.362	82
ASIA countries		
China	0.624	32
Japan	0.786	11
South Korea	0.782	13
North Korea	0.532	52
ASEAN countries		
Singapore	0.925	1
Malaysia	0.893	3
Thailand	0.684	20
Philippines	0.594	37
Indonesia	0.424	70
Lao	0.392	77
Cambodia	0.283	92
Vietnam	0.245	101

Furthermore, based on the data from [Table 3.10], it is visible that several ASEAN countries have higher GCI and global rank in cybersecurity like Singapore, Malaysia, and Thailand than Visegrád countries. In addition, due to the weak cybersecurity capacity, Lao, Cambodia, and Vietnam's positions are quite low. For these reasons mentioned above, these countries need to cooperate together and become one group in order to create common governmental cooperation, strong organization, and cybersecurity capacity building and to solve similar problems in cybersecurity towards global threats.

New key findings on ASEAN cybersecurity strategy cooperation

Several ASEAN countries have started to focus on cybersecurity early and they became the leaders in the same region in processing to develop cybersecurity stability like Malaysia, Indonesia, and Singapore. Malaysia and Indonesia joined in UN Group of Governmental Experts (GGE) meetings to enhance cyber stability and security. In fact, the first ASEAN Telecommunications Ministers Meeting (TELMIN) was hosted in Malaysia in 2001 on the e-ASEAN program to build the e-ASEAN framework agreement. This meeting put out four main objectives such as "*(a) develop, strengthen and enhance the competitiveness of the ICT sector; (b) reduce the digital divide within and amongst ASEAN Member Countries; (c) promote cooperation between the public and private sectors; and (d) develop ASEAN Information Infrastructure*" [310]. In 2011, the ASEAN ICT Masterplan 2015 (AIM2015) was established with an outlook "*Towards an Empowering and Transformational ICT: Creating an Inclusive, Vibrant and Integrated ASEAN*" [310], [311] in order to promote the cooperation between ASEAN Member States (AMS). Five years later, the ASEAN ICT Masterplan 2020 (AIM2020) was adopted in the 15th ASEAN TELMIN with the vision to secure and sustainable digital economy, facilitate transformation; and enable an innovative, inclusive, and integrated ASEAN community [312]. In addition, Singapore has set up the ASEAN cyber capacity program to provide cyber standards and Confidence Building Measures (CBMs) for all nations in the same region [313]. In 2016, Singapore firstly organized meetings between national Ministers on cybersecurity to promote cooperation and develop the standards in ASEAN at the government level [314]. One year later, the ASEAN cybersecurity cooperation strategy was found under Singapore's vice chairmanship of the ASEAN Network security action council with the aim to build the standards, cyber policies, and capacity framework. Moreover, this strategy also focuses on political and security, economic, and socio-cultural community pillars and it follows the framework of TELMIN. Singapore is not only a co-founder nation but also an active member in the cybersecurity capacity-building cooperation in the same region.

It also set up the ASEAN-Singapore Cybersecurity Center of Excellence (ASCCE) in 2018. This center mainly focused on three major pillars such as promoting training and research, training CERTs and enhancing open-source information sharing among CERT in the same area [315]. Furthermore, it was also a leader in the area of cybercrime; for instance, it established a 10$million ASEAN cyber capacity fund to strengthen cybersecurity capabilities for the region [316]. In another hand, AMS also recognizes the demand to protect their cyberspace and ICT infrastructure as quite urgent. Hence, there are four major structures to deal with cybersecurity issues such as the ASEAN Ministerial Meeting on Transnational Crime (AMMTC); ASEAN Telecommunications and IT Ministers Meeting (TELMIN); the ASEAN Regional Forum (ARF), and the ASEAN Senior Officials Meeting on Transnational Crime (SOMTC) [316]. These governmental cooperations were found to fight against transnational crime as well as cybercrime, and cyberterrorism in the area. However, when ASEAN countries work together in cybersecurity, they also face some challenges as follows [313]:

- Inadequate technology, different technology development level, and digital divide between ASEAN members
- Different perception towards cyber issues of policymakers and experts – hard to find common agreement
- Ensuring the regional support for global efforts in cybersecurity instead of fragmentation
- ARF, ASEAN, or GGE mechanisms are not fully successful

Benefits of the transnational approach in the cybersecurity

Japan - ASEAN

As I mentioned above, Japan is a developed country in ICT quite early in the same region. They have good organizations to protect their cyberspace and prevent cyber threats. Besides, Japan also has a good relationship with international nations to fight against global cyber-attacks like the USA and EU. Additionally, Japan and ASEAN have a good relationship in building cybersecurity capacity for ASEAN members. Japan helped ASEAN in creating a draft for information security policy, namely the Critical Information Infrastructure Protection (CIIP) guideline in the ASEAN-Japan Information Security Policy Meeting [317]. Especially, Japan and Singapore signed the agreement on boosting cybersecurity cooperation in 2017 [318]. This agreement with purpose improves cybersecurity awareness, shares the best practices and takes regional capacity-building efforts through policy discussion, information exchanges, and cooperation.

–

Singapore – ASEAN, and others

Beside Japan, Singapore signed another cyber pact with Germany in 2017 to enhance cybersecurity cooperation via information exchange, sharing training, research, and best practices. Since Singapore paid attention to the cyber domain, they expected to build their nation as a developed and secure network country to serve as a center for businesses and attract talents. They established CSA and built a strong partnership with other countries to work in this aspect. In fact, they already signed seven MOUs with France, India, the Netherlands, the UN, the USA, Canada, and Australia [319], [320] to enhance cybersecurity. Particularly, Singapore and the USA work on the Singapore-US cybersecurity Technical Assistance Program for the ASEAN Member States and the USA – ASEAN statement on cybersecurity cooperation [321], [322]. Regarding these declarations, they can improve the regional cybersecurity capacity, infrastructure, and economic development for ASEAN. Likewise, Singapore played an important role among ASEAN members when it composed a formal ASEAN cybersecurity structure to address cyber diplomacy, policy, and operational issues towards cyber-attacks in the region [323].

India - ASEAN

In the 25th anniversary of ASEAN-India Dialogue Relations, India established the Delhi declaration to tighten the relationship between India and ASEAN. In this declaration, India expected to enhance and deepen the ASEAN-India strategic partnership in many aspects such as political-security, economic, socio-cultural, and development collaboration, especially cybersecurity [324], [325]. It emphasized developing the cybersecurity capacity building and policy via applying of ASEAN cybersecurity cooperation strategy, ARF work plan on security in ICTs. Moreover, India also decided to work together in the fighting process against other transnational crimes, cybercrimes, human and drug trafficking, piracy, and armed robbery against ships [326].

EU – ASEAN

EU and Southeast Asia countries have a project namely SEACOOP by the European Commission and the ASEAN Secretariat with the purpose to strengthen ICT cooperation between EU and ten ASEAN countries [258]. This project aimed to identify and analyze the ICT policies and research priorities in AMS in order to decide potential fields for cooperation between ASEAN and the EU Commission. Recently, the EU and ASEAN have a project (Cybersecurity Awareness and Knowledge Systemic High-level Application) – namely YAKSHA [327], [328] in order to build strong cooperation and partnership in the cyber domain in 2018. This project helps the experts in both EU and ASEAN developing new methods to detect malware, collect and analyze vulnerabilities as well as mitigate cyber-threats and enhance the cybersecurity skills of specialists. The EU and ASEAN also focus on strengthening maritime security, terrorism, nuclear weapon, conflict, development of regional cooperative orders, and hybrid threats [329], [330].

Therefore, the EU plays an important role in boosting economic development and improving security cooperation in many fields for ASEAN members.

CONCLUSION

This study provides an overview of the cybersecurity strategy, policies of ASEAN members and other Asia countries. A detailed description of the national cybersecurity strategy of each ASEAN member is given to illustrate the cooperation with international organizations to ensure the safety of critical infrastructure information, strengthen cybersecurity capability building and create the legal framework for cybersecurity. Moreover, consideration is also taken into the role of ASEAN organization for each member in helping to protect their national sovereignty, create general cybersecurity strategy and legal framework foundations. In other words, this organization helps AMS fight against cybercrime, terrorism, cyber-attacks, human trafficking, and the like. This chapter also showed the main important differentiating factor between Asia and EU nations is data protection regulations in Asia countries. It seems like GDPR in the EU but it is more secure because it protects data policy or restricts data, especially in personal data or sensitive data, as well as it does not allow to access data for the third party outside the host like China. Additionally, one new key finding is that the police or military department is responsible for cybercrime units in ASEAN. This type of department organizes the cyber-drill, best practices, or sharing knowledge about cybercrime in order to mitigate and counter against them. Besides, this chapter identified several current challenges in the cooperation of ASEAN members as well as mitigating cyber issues. Furthermore, regarding the transnational cooperation benefits in cybersecurity, ASEAN can take the advantages to improve the cybersecurity capacity building, policy; and protect AMS' cyberspace along with preventing cyber threats. In another way, in this chapter, the author showed that there are several countries with strong cybersecurity capacity in Asia and ASEAN (China, Hong Kong, Japan, South Korea, North Korea, Singapore, and Malaysia) and weak cybersecurity ones (Indonesia, Lao PDR, Cambodia, Vietnam). For the strong cybersecurity nations, they have a good strategy, capacity building, legal framework, and collaboration because of fast approaching in technology, and high cybersecurity awareness; as a result, some of them ranked the top ten of the world about GCI in cybersecurity like Singapore and Malaysia. In contrast, there are several ASEAN members quite weak in capacity building, legal national cybersecurity strategy to defeat cyber-attacks, and response cybersecurity incidents like Cambodia, Lao PDR, and Vietnam. They are hit by a lot of cyber-attacks every year because of lacking experts and technology. As a consequence, they need to build strong cooperation as V4's cooperation to enhance cybersecurity capacity to protect themselves and others in the same region. On the other hand, these countries can self-defense themselves and contribute as one group to ASEAN's development in cybersecurity like Visegrád countries' contribution to EU nations and NATO. Hence, the author strongly accepted that cybersecurity, especially in cybersecurity cooperation in Visegrád countries may be adapted and networked with Asian countries, particularly in Vietnam and its neighbors". Because Vietnam and its neighbors are quite similar to each other in some aspects; for example, small and developing countries, closed geography and same rice agricultural culture, lack of experts and technology, and suffering heavy damage from the war. Thus, I strongly recommend that Lao PDR, Cambodia, Thailand, and Vietnam can cooperate as one group – namely **A4** in cybersecurity aspect like V4 because this group can support the cybersecurity capacity building, enhance the protection of national security, citizens' life, and reduce damage from cyber-attacks for these countries. Likewise, it also helps them to promote a new framework in cybersecurity strategy for ASEAN members.

Chapter Four:

SUGGESTIONS TO APPLY VISEGRAD STRATEGIES FOR ASIAN COUNTRIES (VIETNAM)

4.1. CURRENT CYBERSECURITY CHALLENGES FOR VIETNAM AND ITS NEIGHBORS

Since computer becomes an indispensable thing in individual life and social activities, cyber-attacks are the most serious threat towards politics, the economy, the military, and national security for all countries. Nowadays, hackers or cyber-crimes are more complicated. They used many types of cyber-attacks to penetrate the systems; steal sensitive or personal information for financial or political benefits; destroy the country's cyber defense. With the boosting of technology, they can take advantages to easily attack many countries at the same time, especially in developing countries or less developed technology nations with a lot of security vulnerabilities such as ASIA or ASEAN nations. In fact, ASEAN countries that have non–state cooperation and a lot of differences in perception of cybersecurity, cyber capability; as well as a big gap of digital level among members. These are honeypots for hackers to take profits. Moreover, there is a lack of trust or transparency in sharing cyber incident knowledge or threats amongst ASEAN nations. Therefore, it is hard to cooperate in order to detect, prevent, protect or investigate cyber-attacks in time. Particularly, several countries with developing an economic system such as Laos, Cambodia, and Vietnam they have less digital development in comparison with the others; as a result, it is extremely difficult for them to make decision or solutions towards cyber-attacks or cyber incidents. Hence, they are on the top of most countries under cyber-attacks because of lacking legal national cybersecurity strategy, experts, and technology.

Proposal for cybersecurity strategies for Vietnam

Regarding the author's research, there are several proposals for cybersecurity strategies for Vietnam, as follows:

Option 1: Singapore cybersecurity strategy can be adapted in Vietnam
Firstly, Singapore is one of ASEAN countries and a small country with its population of approximately 5.792 million people in 2018; nonetheless, the technology development's speed is extremely high and it quickly becomes not

only a massive technology hub but also a finance center in the world. In fact, its global cybersecurity index (GCI) in 2017 was the first rank in the world with a score of 0.925 [112]. Due to the aim of building the Smart Nation infrastructure, Singapore spent a lot of budgets of 1% of GDP on R&D for scientific and technology research [331]. Meanwhile, Vietnam is a developing country with a crowded population and high-speed approach in technology; as a result, the Vietnam government needs to build concrete and resilient infrastructure systems for government and officials similar to Singapore [231]. For instance, they may create the government networks for e-government and e-business to keep important services and let the participation of all stakeholders – government, private sectors, security community in order to ensure the safety of sharing information among them. Besides, the government desires to implement several programs or projects to protect critical infrastructure information from cyber threats. Additionally, enhancing cyber capability and improving the legal framework to address cyber threats are also urgent requirements. For instance, Vietnam government may establish more cyber laws, Acts or Decrees, especially national cybersecurity strategy which declare responsibilities and function of agencies, operators, private sectors and public sectors to safeguard their system, networks and protect government system and citizens as well. Secondly, Vietnamese requires to encourage the cooperation of private sectors like VNISA, VSISA, VIA, VEA, VAIP, BKAV, white hacker teams, and the like with public sectors (VNCERT) to share cybersecurity bills, exercises, or experiences in order to increase cybersecurity awareness for citizens, officials, and organizations towards cyber-threats; and know-how to protect sensitive data or information. Indeed, in 2017, according to the Vietnamese Ministry of Information and Communication (MIC), Vietnam ranked 16th out of 20 countries in using the Internet in ASIA with approximately 53% Internet users over the population; however, based on the International Telecommunication Union (ITU) and Global Cybersecurity Index (GCI) in the same year, Vietnam ranked 101th of 193 countries in network security [332]. For this reason, the role of education in cybersecurity is an essential and urgent requirement to raise the safety and security information's awareness for every individual or organization through training, contests, or educational programs in school or universities. Thirdly, Vietnam needs to build more institutions, training centers, or universities to train or educate the experts in the cybersecurity aspect. In 2017, according to the Vietnamese Department of Information Security and Communication, there was 8 institutions or universities which recruited the students in information security and 953 IT security engineers in the whole nation [333]. Finally, the Vietnamese government requires to expand the international cooperation in cybersecurity with several developed countries in the same region like Singapore, Malaysia, Japan or another region like the EU and the USA.

Option 2: Visegrád countries' cybersecurity strategies can be adapted towards Vietnam and other neighboring countries.

In ASEAN nations, Vietnam, Thailand, Lao PDR, and Cambodia (**A4**) are the small and developing countries that have quite weak cybersecurity in the same region. However, they have several similar aspects such as social, geographical position, agriculture and rice cultivation, and historical development.

Regional and religious specifications

They are part of a peninsula of southeastern Asia (Indochinese peninsula-Indochina) which includes Myanmar, Cambodia, Lao PDR, Malaysia, Thailand, and Vietnam [334], [335]. The common of Indochina's physical environment is mainly mountainous. Particularly, the climate in these countries is a monsoon tropical climate type (raining in summer and drying in the winter). For instance, Cambodia experiences a tropical savanna climate while Vietnam and Laos have two different climates between the north (humid subtropical climate) and the south (tropical rainy climate) [336]. Besides, the most common religions in these countries are Hindu and Buddhist and these countries are related to Chinese's cultural areas like language and shared religious factors (Confucianism, Buddhism, and ancestor veneration). Moreover, these countries have the same river called Mekong River – a trans-boundary river which starts from China to Myanmar, Laos, Thailand, Cambodia and ends in Vietnam. They use this river's resources as a natural resource for developing the water rice cultivation, as a result, they have the same agriculture culture for developing the economy. Although they have natural resources to enable potential economic development, they are lack of technology and budgets to take benefits from them because of suffering a long time with the war in the past. Therefore, they still remain the world's poorest countries.

Historical factors

Vietnam, Laos, Cambodia also had nearly the same history when they were colonies of several strong power countries like the USA, France, Japan, and China. For instance, France conquered three Indochina countries as Vietnam about 83 years, Laos PDR 53 years, and Cambodia 82 years [337], [338], [339]. Moreover, Vietnam was the colony of the USA (1948-1975), China (111 BC – 938 AD) [340]; and these countries were also conquered by Japan from (1940-1945). After a long period in the war, three countries suffered extreme damage to the economy, infrastructure, and citizens' life. Even though Thailand is one of the Indochina countries, it is the only nation in Southeast Asia that avoids the conquest by any European countries because the French and the British considered as a neutral region to prevent the conflicts between their colonies [341]. Since their independence, these countries have focused on developing the economy, military, education, and so on.

Cybersecurity states in A4 (Thailand, Lao PDR, Cambodia, and Vietnam)
According to GCI 2017 [112], Thailand and Lao PDR were on the same "Maturing" stage with Visegrád countries (Czech Republic, Poland, Hungary, and Slovakia) in cybersecurity, meanwhile, Cambodia and Vietnam were still in the "Initiating" stage countries. In fact, Thailand was the highest nation in comparison with V4 and its neighbors in GCI 2017- ranked the 20th, Czech Republic 35th, Poland 33rd, Hungary 51st Slovakia 82nd, Lao PDR 77th, Cambodia 92nd, and Vietnam 101st, respectively. Thus, Thailand can be the leader of the **A4** group in cybersecurity cooperation like Poland's role in V4 cooperation. Currently, Thailand, Lao PDR, and Vietnam already have standalone cybersecurity laws but Cambodia has a draft in cybercrime law and it has not been affected yet [342]. Nevertheless, these countries signed MoU among CERTs to enhance cybersecurity against cyber – attacks. Hence, when the cooperation amongst these countries is established, this may support cyber policy development, enhance capacity building and facilitate operational issues in preventing cyber-attacks and promoting cybersecurity for these nations and ASEAN in general.

International cooperation project (if any)

Child online protection
Nowadays, there are many types of crimes related to children such as child pornography, sexual exploitation, child trafficking, child labor, forced marriage, prostitution, and so on [343]. Particularly, these crimes usually occur in the developing countries in ASEAN such as Lao PDR, Vietnam, Cambodia, Thailand, and Myanmar. In fact, in Thailand, children are traded from Cambodia, Lao PDR, and Myanmar with the aims for labor trafficking, sexual exploitation, and forced begging [343] because it is quite easy to enter Thailand via the border by various means of transportation from these countries. In addition, the United Nations Office on Drugs and Crime (UNODC) reported that Vietnamese children trafficking victims were found in neighbor countries. According to the International Telecommunication Union (ITU), Child online protection is a global issue that needs the cooperation of all nations at the international level. It also clarifies five major keys to protect and develop child online protection such as "legal measures, technical and procedural measures, organizational structures, capacity building, and international cooperation" [344]. ITU started the Child Online Protection (COP) Initiative in 2008 within the Global cybersecurity agenda framework [345]. It supports the Member States, especially in developing countries to promote and deploy guidelines for the COP initiative. At present, Thailand, Cambodia, and Vietnam are the members of ITU except for Lao PDR; as a consequence, with the cooperation

of A4, this cooperation helps not only Lao PDR but also four nations in improving public awareness; sharing best practices, tools, and resources to adapt in each country; clarifying policies, risks and vulnerabilities; and enhancing the capacity building in protecting child online [346].

Human trafficking

Human trafficking refers to three main types such as sexual exploitation, labor exploitation, and organ trafficking [347], [348]. According to UNODC, the human trafficking victims found in East Asia and the Pacific more than 85 percent, 6 percent from South Asia. Besides, trafficked victims from Indonesia, the Philippines, and Vietnam were mostly found in Malaysia while the human trafficking people from Cambodia, Lao PDR, and Myanmar were recognized in Thailand [349]. However, in 2015, on the 27th ASEAN SUMMIT, it established the ASEAN Convention Against Trafficking in Persons, Especially Women and Children to combat trafficking in person, especially women and children (ACTIP) [350]. This convention requires at least six members of ASEAN countries to ratify it in order to go into effect. Until now, Thailand, Lao PDR, Singapore, Cambodia, Vietnam, and Myanmar have ratified the convention against human trafficking [351],[352],[353],[354],[355]. It is visible that four countries of A4 have ratified the convention to fight against human trafficking, as a consequence, this can increase public awareness of trafficking in persons, people smuggling, and transnational crime for each other and other ASEAN countries in the same region. In another hand, A4 countries are also members of the International Criminal Police Organization (INTERPOL) – an inter-governmental organization with 194 members located in Singapore [356]. This organization gives technical and operational support to help police amongst members in sharing and accessing data on crimes and criminals. In addition, this organization mainly focuses on three crime's programs such as counter-terrorism, cybercrime, and organized and emerging crime. Human trafficking is one kind of international organized crime that Interpol helps its members to counter against. Hence, with the cooperation of A4, it can help these countries prevent human trafficking themselves and it can contribute to Interpol operations and the other members in the same region.

Economics

As mentioned above, Thailand, Laos, Cambodia, and Vietnam are Indochina countries and they have the same water rice cultivation culture. Moreover, their large contribution to the economy is based on the export of agricultural products. For example, in 2017, the Gross Domestic Products (GDP) per capita of these countries was Thailand 6,125$ USD, Lao PDR 1730.40$ USD, Cambodia 1,379.34$ USD, and Vietnam 1834.65$ USD, respectively [357], [358], [359], [360]. Remarkably, Thailand ranked the 2nd after Indonesia, while

Vietnam ranked the 6th, Lao PDR ranked 9th, and Cambodia ranked the 8th about GDP in 2018 [Figure 4.1]. At present, ASEAN nations have the Free Trade Area (FTA) with some countries such as China, Japan, the Republic of Korea, Australia, New Zealand, and India [361]; as a consequence, the A4 group may take the advantages of this agreement to boost their competitiveness, trade development, expand cross-border cooperation with the other nations. Besides, Vietnam began a new step in dealing with a free trade agreement between Vietnam and the EU (EVFTA) which is in the process of final ratification from European Council before it comes into force [362]. With this agreement, it may bring a lot of benefits for both Vietnam and EU members. In fact, Vietnam is a potential trade partner of EU after Singapore based on heavily exporting products and lower wages labor's price in some aspects such as mobile and electronic products, footwear, textiles and clothing, coffee, rice, seafood, and furniture. In another hand, if this agreement is established, it can open up big opportunities not only for Vietnam but also for A4 group in reducing tariffs on goods, increasing trade competitiveness in the same region, developing economy growth, expanding the market to a new area, and promoting their position in the global.

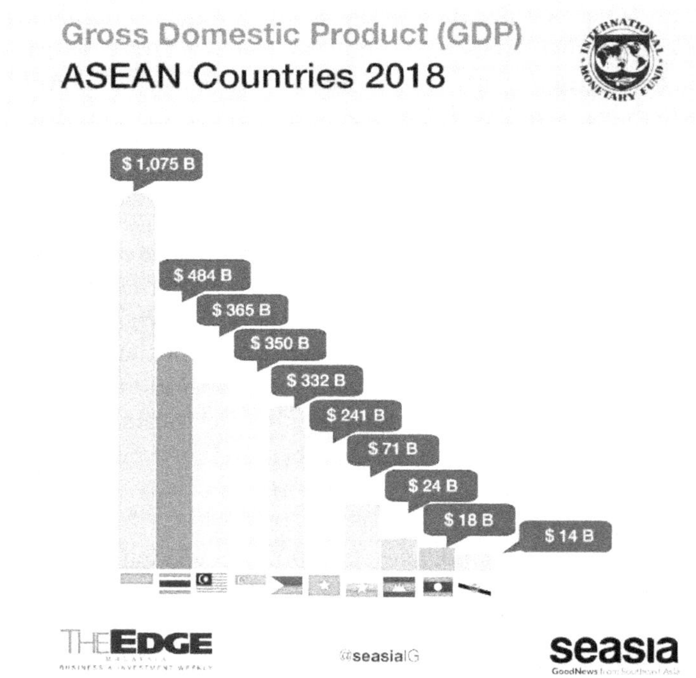

Figure 4.1: Ranking of GDP Per Capita of Southeast Asian Countries [363]

CONCLUSION

This research provided the answers for six questions, as follows:

1. What are the current cyber-threats?
2. Are they dangerous threats to Vietnam?
3. What can Viet Nam do to mitigate cyber-threats?
4. How can Viet Nam cooperate with international organizations to solve these risks?
5. What are the benefits of Visegrád strategies in defense towards cyber-attacks with other countries?
6. Which cybersecurity strategies are found suitable for Vietnam in terms of adaptation?

In order to achieve the reseach purposes, the thesis generally overviewed the types of cybercrime (man-made attacks and machine-made attacks) and cyberwarfare towards the national security, the negative and different impacts of them on the security at the government level and citizens' life; especially in the economy, finance and information infrastructure system of a nation. Moreover, this thesis particularly showed the bilateral and multilateral cooperation among Asian countries in security and economy, as well as in trade, economy, military, energy, peace, friendship, and diplomacy for ASEAN nations [364].

Moreover, it gives significant differences in cybersecurity cooperation between Asia and the EU. For example, in Asia countries, cybersecurity cooperation mainly focused on sharing information and knowledge to prevent the cyber-attacks towards economy via several private sectors in finance and intelligence because of non-state political connection [365]. In contrast, the EU nations have the same legal framework, standards, strategies, and regulations; therefore, their cybersecurity cooperation not only concentrates on the safety of politics but also on the security of cyberspace to reduce the damage and protect their national sovereignty or national security. In addition, data protection regulations or data policy in several Asian countries are more secure than GDPR in the EU because they restrict the third party outside the host country to access the data. Furthermore, it made clear the cybersecurity cooperation amongst V4 countries considered as one nation's power in the center of EU in order to strengthen national stability, decrease the cyber-threats, enhance the relationship and improve the cybersecurity, cyber defense or other future challenges between EU, NATO and other organizations [366]. Likewise, there are several organizations like ENISA, NATO, the Three Seas Initiative, and E3PR which offer general legal frameworks (GDPR, NIST 800-53, NIS directive, Digital Single Market Initiative, and CPPP) in cybersecurity cooperation strategies for EU countries.

This study has presented an overview of the ASEAN nation's cybersecurity strategy and its current cyber challenges. Simultaneously, it has revealed weaknesses in response to cyber incidents and low awareness about the importance of national information cybersecurity of some countries because of the inadequacy of cybersecurity cooperation with the others in the same region.

In the first chapter, this research identifies two major types of cybercrimes such as machine-made and man-made attack in order to have a general outlook of these cybercrimes and their influences towards the government, citizen's life, and the safety of a country. In addition, one important key factor is that this research also reveals the differences between cybercrime and cyber-warfare; for instance, cybercrime mainly focuses on economy and finance while cyberwarfare aims at politics, critical infrastructure, and security of nation or citizens. In another hand, this research also emphasizes the bilateral, trilateral, or multilateral cooperation in cybersecurity and public security between Asian and ASEAN countries with the USA, China, Russia, and European countries. Due to a variety of individual perceptions about cybersecurity, lack of cybersecurity capacity building between Asia and Europe, Asian countries' cooperation is mainly focusing on the economy, military, and diplomacy. In fact, it indicates several Asian organizations' cooperation for financial security such as FS-ISAC and HTCIA.

Summarizing on the second chapter, we analyzed and compared the cybersecurity strategies of Visegrád countries through their documents, cyber-laws, and so on. We clarified the differences amongst V4 countries in protecting national security strategies, institutional backgrounds, and political plans. We figured out the major difficulty of these countries is the lack of experts in the public and private sectors. In another hand, we recognized that they have the same main aims to ensure national security level and contribute to cybersecurity agendas of the EU and NATO. In addition, we found out the common security threats of Visegrád countries towards their national security level such as terrorism, cyber-attacks, international immigration, regional conflicts, and transnational crimes, interruption of supplies of raw materials or energy, and natural disasters. As a result, we determined the V4 cooperation not only helps themselves but also promotes EU and NATO in security structure in cybersecurity, cyber-defense more effectively. Moreover, this cooperation enriches the power of V4 nations in supporting military capabilities, armed forces, cyber-defense, energy supplement, cybersecurity as a nation in the EU [367].

Regarding on the information in chapter three, we introduced several particular East Asia and the South East Asian countries' cybersecurity situation. Based on the main challenges, aims and cybersecurity capacity in protecting cyberspace, we defined two main groups like several strong nations in cybersecurity's capacity building and the weak ones. we considered that the strong cybersecurity capacity nations including China, South Korea, North Korea, Japan, Hong Kong, Singapore, and Malaysia have an early cybersecurity strategy, strong cybersecurity policy as well as cyber-laws, legal framework to handle the cyber-threats. In contrast, we pointed out the important problems of the weak cybersecurity capacity-building countries (the Philippines, Thailand, Cambodia, Indonesia, Cambodia, and Vietnam) are the inadequacy of technology, experts, and budgets to build strong cybersecurity strategy and capacity-building in cybersecurity and cyber-defense. Otherwise, we distinguished that Singapore's cybersecurity strategy can be used for Vietnam's legal framework.

In addition, in the chapter four, we analyzed the common history, geography, and culture of Vietnam and its neighbors (Lao PDR, Cambodia, and Thailand). Then, we found that they are not only similar to culture, suffering heavy damage from the war, history but also they are lack technology and experts in cybersecurity aspect. Moreover, they are less digitally developed countries, with low cybersecurity awareness, capacity building, and high challenges in global cyber-threats. Consequently, we highly believe that the cooperation of A4 nations can enhance their cybersecurity capabilities, mitigate the damage from cyber-attacks. Additionally, we analyzed the cooperation of Visegrád countries and I figured out these countries were also the same in some aspects like A4 group such as small countries, close geography, history, culture, security problems, and so on. This cooperation supported each nation in the V4 group in many aspects; therefore, we highly believe that the cooperation amongst Vietnam and its neighbors not only supports each nation but also contributes to ASEAN members' development in many areas. Fortunately, Vietnam is one country that has cooperated with the Czech Republic in cybersecurity. As a result, when Vietnam and its neighbors cooperate together, they can share technology, best practices in cybersecurity, or cyber-defense more conveniently and effectively. After all, we strongly agree that the cybersecurity cooperation of V4 can be applied and networked with Asian countries, especially Vietnam and its neighbors.

Likewise, the main problems in the cybersecurity of Vietnam and its neighbors are reported. Due to a lack of technology, experts, and legal national cybersecurity strategy; Vietnam and its neighbors are the targets of a lot of cyber-attacks every year. In addition, sharing cyber incidents or best practices is also a major difficulty amongst the ASEAN members because they are non-state or federal cooperation, different perceptions about cybersecurity, and lack of trust; therefore, it is highly hard to give the decision in time towards the cyber-attacks. Besides, ASEAN nations almost are developing countries; consequently, their infrastructure is quite low to approach the new technology in order to prevent, protect or mitigate the cyber-attacks. On the other hand, the main key point is that we propose two options for enhancing cybersecurity strategy for Vietnam; for instance, Singapore cybersecurity strategy may be adapted for Vietnam's legal framework and Visegrád countries' cybersecurity strategies also can be used for Vietnam and its neighbors in cybersecurity cooperation.

REFERENCES

[1] Lopez Research, "An Introduction to the Internet of Things (IoT)," *Lopez Res. Llc*, vol. Part 1. of, no. November, pp. 1–6, 2013.

[2] F. D. Janos and N. H. P. Dai, "Security concerns towards security operations centers," *SACI 2018 - IEEE 12th Int. Symp. Appl. Comput. Intell. Informatics, Proc.*, pp. 273–278, 2018.

[3] R. Z. Nguyen Huu Phuoc Dai, *INTERNATIONAL CONFERENCE ON APPLIED INTERNET AND INFORMATION TECHNOLOGY.* .

[4] "Vietnam and threats of cyber attacks." [Online]. Available: http://english.vietnamnet.vn/fms/science-it/155532/vietnam-and-the-threat-of-cyber-attacks.html.

[5] "Critical infrastructure targeted by cyber attacks." [Online]. Available: https://www.infosecurity-magazine.com/news/40-of-ics-critical-infrastructure/.

[6] "CMC Infosec says malware used to attack Noi Bai Airport." [Online]. Available: http://english.vietnamnet.vn/fms/science-it/161772/cmc-infosec-says-malware-used-to-attack-noi-bai-airport.html.

[7] "chinese attacked 1000 vn website." [Online]. Available: http://tuoitrenews.vn/society/28449/chinese-hackers-attack-1000-vietnamese-websites-in-two-days.

[8] L. Almann and J. J. Kelly, "CRS report for Congress - Economic Impact Cyber-Attacks," *Policy Rev.*, p. 39+, 2008.

[9] J. Hua and S. Bapna, "The economic impact of cyber terrorism," *J. Strateg. Inf. Syst.*, vol. 22, no. 2, pp. 175–186, 2013.

[10] S. Gour, "" Cyber Attacks : An impact on Economy to an organization ,"" vol. 4, no. 9, pp. 937–940, 2014.

[11] H. Porteous, "or a Game Changer ?," 2010.

[12] "The countries are attacked by Wannacry ransomware." [Online]. Available: https://blog.avast.com/ransomware-that-infected-telefonica-and-nhs-hospitals-is-spreading-aggressively-with-over-50000-attacks-so-far-today.

[13] P. J.Smith, "Transnational security threats and state survival: A role for the military?," *Parameters*, pp. 77–91, 2000.

[14] M. Wesley, ""Transnational Crime and Security Threats in Asia,"" vol. 2, no. December 2000, pp. 1–15, 2000.

[15] "Evidence of computer vandalism in MOE website in VN." [Online]. Available: http://news.zing.vn/thanh-nien-19-tuoi-lap-website-gia-chiem-doat-hon-140-trieu-dong-post758045.html.

[16] J. Cockayne and C. Mikulaschek, "Major Terror Attacks in Bangladesh Transnational Security Threats Challenging Bangladesh."

[17] United Nations Office on Drugs and Crime (UNODC), "Transnational Organized Crime – The Globalized Illegal Economy," 2009.

[18] D. Kar, "Transnational Crime and the Developing World," no. June 2016, 2017.

[19] G. Wacker, "Security Cooperation in East Asia. Structures, Trends and Limitations," no. May, 2015.

[20] S. Asia, *South-East Asia*, vol. 2005, no. September. 2006.

[21] "U.S. Relations With Japan Share," 2018. [Online]. Available: https://www.state.gov/r/pa/ei/bgn/4142.htm.

[22] "Japan-United States of America Relations." [Online]. Available: https://www.mofa.go.jp/region/n-america/us/security/arrange.html.

[23] "USA - South Korea cooperation," 2016. [Online]. Available: http://countrystudies.us/south-korea/76.htm.

[24] "U.S. Relations With Australia Share," 2018. [Online]. Available: https://www.state.gov/r/pa/ei/bgn/2698.htm.

[25] Government of Australia and Government of the United States of America, "Australia–United States Free Trade Agreement," 2004. [Online]. Available: https://www.austrade.gov.au/Australian/Export/Free-Trade-Agreements/AUSFTA.

[26] "United States and Australia Intensify Cooperation on Digital Trade," 2018. [Online]. Available: https://ustr.gov/about-us/policy-offices/press-office/press-releases/2018/february/united-states-and-australia.

[27] "Philippines–United States relations," 2018. [Online]. Available: https://dfa.gov.ph/dfa-news/dfa-releasesupdate/2695-philippine-united-states-bilateral-relations-fact-sheet.

[28] "Thailan and US in security cooperation." [Online]. Available: https://www.state.gov/r/pa/ei/bgn/2814.htm.

[29] "US-Russia cooperation," 2017. [Online]. Available: https://www.americansecurityproject.org/us-russia-relationship/us-russia-cooperation/.

[30] "USA and Russian relations." [Online]. Available: https://www.csis.org/programs/russia-and-eurasia-program/archives/us-russian-relations.

[31] M. W. Märta Carlsson, Susanne Oxenstierna, "Russian and china cooperation," 2015. .

[32] "Deepened Strategic Partnership of Coordination between China and Russia." [Online]. Available: http://www.ciis.org.cn/english/2011-08/12/content_4404233.htm.

[33] E. Sinkkonen, "China-Russia Security Cooperation: Geopolitical Signalling with Limits," 2018. [Online]. Available: https://isnblog.ethz.ch/internation-

al-relations/china-russia-security-cooperation-geopolitical-signalling-with-limits.

[34] R. M. David M Malone, "India and China: Conflict and Cooperation," 2010. .

[35] A. R. Nalpathamkalam, "Cooperation without trust: India-China relations today," *2012*, 2012.

[36] "US-China Relations: Competition or Cooperation?," 2017. [Online]. Available: http://www.globaltrademag.com/in-the-news/us-china-relations-competition-cooperation.

[37] D. D. Finkelstein, "The Military Dimensions of U.S. - China Security Cooperation: Retrospective and Future Prospects," 2010. .

[38] "Comparing Cyber-Relations: Russia, China, and the U.S," 2016. [Online]. Available: http://natoassociation.ca/comparing-cyber-relations-russia-china-and-the-u-s/.

[39] "Treaty of Friendship, Co-operation and Mutual Assistance Between the People's Republic of China and the Democratic People's Republic of Korea," 1961. [Online]. Available: https://www.marxists.org/subject/china/documents/china_dprk.htm.

[40] "The China–North Korea Relationship," 2019. [Online]. Available: https://www.cfr.org/backgrounder/china-north-korea-relationship.

[41] Mark Cartwright, "Ancient Japanese & Chinese Relations," 2017. [Online]. Available: https://www.ancient.eu/article/1085/ancient-japanese--chinese-relations/.

[42] R. H. Hanns Gunther Hilpert, "Hilpert and Haak, Japan and China: Cooperation, Competition and Conflict, 2002," 2003. [Online]. Available: https://china.usc.edu/hilpert-and-haak-japan-and-china-cooperation-competition-and-conflict-2002.

[43] "Japan-India Relations," 2018. [Online]. Available: https://www.mofa.go.jp/region/asia-paci/india/data.html.

[44] Indian Embassy, "India-Japan relationship," 2016.

[45] "VN sign MOU with Czech." [Online]. Available: http://english.vietnamnet.vn/fms/science-it/176606/vietnam--czech-firms-sign-mou-on-cyber-security-cooperation.html.

[46] "High technology crime investigation association." [Online]. Available: https://htcia.org/.

[47] "Financial Services Information Sharing and Analysis Center." [Online]. Available: https://www.fsisac.com/.

[48] M. D. G. and S. W. Brenner, "The emerging consensus on criminal conduct in cyberspace," *World*.

[49] P. Kleve, R. De Mulder, and K. Van Noortwijk, "The definition of ICT Crime," *Comput. Law Secur. Rev.*, vol. 27, no. 2, pp. 162–167, 2011.

[50] National Crime Prevention Council, "Cybercrimes," *Bur. Justice Assist.*, pp. 1–4, 2012.

[51] N. LEENA, “Cyber Crime Effecting E-commerce Technology,” *Orient. J. Comput. Sci. Technol.*, vol. 4, no. 1, pp. 209–212, 2011.

[52] M. E. Kabay, “A Brief History of Computer Crime: An Introduction for Students,” *Security*, pp. 1–51, 2008.

[53] N. Kamath and C. Secretary, “Cyber crime,” pp. 54–180, 2003.

[54] W. Mitchell, L. Review, R. J. Mcgillivray, and S. C. Lieske, “Webjacking,” vol. 27, no. 3, 2001.

[55] “Web jacking statistics 2017.” [Online]. Available: https://www.go-gulf.com/blog/cyber-crime/.

[56] A. Khatri, R. Savla, A. Malde, and D. Pawade, “Cybercrimes and Attacks for Data Access , Online Transaction with Overview Of Cyber Security Acts,” no. April, pp. 723–725, 2016.

[57] L. V. A, “Salami Theft-Major Threat To Information Security,” pp. 2319–2321, 2015.

[58] “Evidence of Salami attack.” [Online]. Available: https://www.wired.com/2008/05/man-allegedly-b/.

[59] R. Wortley, S. Smallbone, and U. Services, *Child pornography on the Internet*, vol. 18, no. 41. 2006.

[60] M. Taylor and E. Quayle, “Child Pornography: An Internet Crime,” no. May, p. 236, 2003.

[61] P. F. Fagan and D. Ph, “The Effects of Pornography on Individuals, Marriage, Family, and Community,” *Addiction*, no. December, pp. 1–26, 2009.

[62] B. P. Goff, “Computer Vandalism , Fraud and Other Forms of Thievery Don ’ t be a Victim We ’ re retired . Why would thieves bother with us ? We Are Prime Targets !,” 2012.

[63] S. L. Doina Bein, Wolfgang W.Bein, Vasu Jolly, “Guarding against Web spoofing and phishing attacks,” pp. 7–10, 1903.

[64] M. Nimmo, “Fraud and money laundering,” no. 31, pp. 1–16, 2007.

[65] I. N. Confidence, “Money laundering,” no. July, 2005.

[66] P. Reuter and E. M. Truman, “Money Laundering: Methods and Markets,” *Chas. Dirty Money Fight against money Laund.*, pp. 25–43, 2003.

[67] J. L. McCURDY, “Computer Crimes,” *Am. Crim. L. Rev.*, vol. 47, pp. 287–1341, 2010.

[68] T. P. Bhatla, V. Prabhu, and A. Dua, “Understanding Credit Card Frauds,” *Cards Bus. Rev.*, vol. 1, no. 6, pp. 1–15, 2003.

[69] I. Crime, “Internet Crime Report - 2010,” p. 24, 2010.

[70] C. Reilly and N. Smith, “Internet Gambling : An Emerging Field of Research,” p. 5, 2013.

[71] C. McFarland, F. Paget, and R. Samani, “Jackpot! Money Laundering Through Online Gambling,” *McAfee Labs Exec. Summ.*, 2014.

[72] K. Young, “Understanding Compulsive Online Gambling and Treatment for Addicts The Seductive Nature of Online Gambling.”

[73] "What did the researchers find? •," no. July 2005, pp. 1–18, 2007.
[74] Buraeu of Justice Assistance, "PREVENTING IDENTITY THEFT : a GUIDE for CONSUMERS," 2005.
[75] "FRAUD / HACKING E-mail bomb suspect arrested ~ The system has been brought to," no. February, p. 1996, 1996.
[76] "Email bombing is making a quick comeback." [Online]. Available: https://smartermsp.com/emailing-bombing-old-new/.
[77] S. Hinduja and J. W. Patchin, "Cyberbullying: An Exploratory Analysis of Factors Related to Offending and Victimization," *Deviant Behav.*, vol. 29, no. 2, pp. 129–156, 2008.
[78] S. V. Sari and F. Camadan, "The new face of violence tendency:Cyber bullying perpetrators and their victims," *Comput. Human Behav.*, vol. 59, pp. 317–326, 2016.
[79] S. Hinduja and J. W. Patchin, "Cyberbullying: identification, prevention & response," *Cyberbullying Res. Cent.*, no. October, pp. 1–9, 2014.
[80] "Posted sex photos, videos on the facebook." [Online]. Available: http://www.baomoi.com/ke-tung-clip-sex-voi-nu-sinh-lop-11-quang-binh-nhan-cai-ket-dang/c/21974456.epi.
[81] T. Morkel, J. H. P. Eloff, and M. S. Olivier, "An Overview of Image Steganography," *Inf. Comput. Secur. Archit. Res. Gr.*, vol. 83, no. July, pp. 51–107, 2005.
[82] I. J. Cox, M. L. Miller, J. A. Bloom, J. Fridrich, and T. Kalker, "Special Issue on Steganography and Digital Watermarking," *Network*, vol. 153, no. 3, pp. 2005–2006, 2006.
[83] "computer vandalism." [Online]. Available: https://usa.kaspersky.com/internet-security-center/threats/computer-vandalism#.WNl666IlFdg.
[84] G. Weimann, "Cyberterrorism How Real Is the Threat?," *United States Inst. Peace*, no. Special Report 119, pp. 1–12, 2004.
[85] S. M. Furnell and M. J. Warren, "Computer hacking and cyber terrorism: the real threats in the new millennium?," *Comput. Secur.*, vol. 18, no. 1, pp. 28–34, 1999.
[86] E. A. Schulman, "The Elephant in the Room," *Headache*, vol. 50, no. 1, pp. 3–4, 2010.
[87] R. E. Fund, "2 0 1 5," pp. 5–8.
[88] J. Achkoski and M. Dojchinovski, "Cyber Terrorism and Cyber Crime – Threats for Cyber Security," *Proc. First Annu. Int. Sci. Conf.*, 2012.
[89] "North Korea is a bigger cyber-attack threat than Russia." [Online]. Available: https://www.theguardian.com/technology/2018/feb/26/north-korea-cyber-attack-threat-russia.
[90] "Exclusive: North Korea's Unit 180, the cyber warfare cell that worries the West."
[91] R. C. Parks and D. P. Duggan, "Principles of cyberwarfare," *IEEE Secur. Priv.*, vol. 9, no. 5, pp. 30–35, 2011.
[92] C. Czosseck, R. Ottis, and K. Ziolkowski, *4th International Conference on Cyber Confl ict*. 2012.

[93] C. G. Billo and W. Chang, "Cyber Warfare An Analysis of the means and motivations of selected nation states.," *Office*, no. December, p. 142, 2004.
[94] Moran, *A Cyber Early Warning Model*. 2009.
[95] I. Bernik, *Cybercrime and Cyberwarfare*. 2014.
[96] D. Rubenstein, "Nation State Cyber Espionage and its Impacts," pp. 1–11, 2014.
[97] H. Dalziel, "The four amigos : Stuxnet , Flame , Gauss and DuQu," *General Hacking Posts*. [Online]. Available: https://www.concise-courses.com/stuxnet-flame-gauss-duqu/.
[98] Major Shcaap, Arie J., "Cyber Warfare Operations: Development and Use Under International Law," *Cardozo J. Int. Comp. Law*, vol. 64, pp. 121–172, 2009.
[99] M. Watney, "Challenges pertaining to cyber war under international law," *2014 3rd Int. Conf. Cyber Secur. Cyber Warf. Digit. Forensic, CyberSec 2014*, pp. 1–5, 2014.
[100] W. Post and E. Nakashima, "US Target of Massive Cyber- Espionage Campaign," 2013.
[101] "Cyber sabotage." [Online]. Available: https://www.defensetech.org/2008/02/06/cyber-sabotage/.
[102] J. Stone, "Cyber War Will Take Place!," *J. Strateg. Stud.*, vol. 36, no. 1, pp. 101–108, 2013.
[103] "Internet Security Threat Report," vol. 21, no. April, 2016.
[104] L. Gallon and P. Owezarski, "Network Security and DoS Attacks 0.," no. April, pp. 1–24, 2005.
[105] Q. Gu and P. Liu, "Denial of Service Attacks," *Handb. Comput. Networks*, vol. 3, pp. 454–468, 2012.
[106] D. Dittrich, "The DoS Project's 'trinoo' distributed denial of service attack tool." [Online]. Available: https://staff.washington.edu/dittrich/misc/trinoo.analysis.txt.
[107] "Ddos attack in Russian banks."
[108] "Fake news in the V4: governments are often part of the problem," 2018. [Online]. Available: http://visegradinfo.eu/index.php/80-articles/564-fake-news-in-the-v4-governments-are-often-part-of-the-problem.
[109] "Fake news and disinformation." [Online]. Available: https://ec.europa.eu/digital-single-market/en/fake-news-disinformation.
[110] "Disinformation and the cyber threat to digital democracies." [Online]. Available: https://www.governmenteuropa.eu/cyber-threat-to-digital-democracies/86994/.
[111] S. Morgan, "Fake news, disinformation, manipulation and online tactics to undermine democracy," *J. Cyber Policy*, vol. 3, no. 1, pp. 39–43, 2018.
[112] ITU, *Global Cybersecurity Index & Cyberwellness Profiles 2017*. 2017.
[113] "The EU General Data Protection Regulation (GDPR)." [Online]. Available: https://eugdpr.org/.
[114] "General data protection Regulation (GDPR)," 2018. [Online]. Available: https://gdpr-info.eu/.

[115] "Japan and EU agree on terms of reciprocal adequacy for data transfer." [Online]. Available: https://www.alstonprivacy.com/japan-and-eu-agree-on-terms-of-reciprocal-adequacy-for-data-transfers/.
[116] "The lack of cybersecurity capacity building in Asian countries." [Online]. Available: https://www.thegfce.com/news/news/2016/06/20/the-lack-of-cyber-security-capacity-building-frameworks-in-asia.
[117] "ASEAN cybersecurity profile." [Online]. Available: https://jsis.washington.edu/news/asean-cybersecurity-profile-finding-path-resilient-regime/.
[118] "Cybersecurity capacity portal." [Online]. Available: https://www.sbs.ox.ac.uk/cybersecurity-capacity/explore/capacity_dimensions.
[119] T. O. Assessment, *Challenges and risks for the EU.* .
[120] BSA The Software Alliance, "EU Cybersecurity Dashboard: A Path to a Secure European Cyberspace," *BSA Softw. Alliance*, 2016.
[121] I. Technology and A. C. Dashboard, "COUNTRY : VIETNAM," pp. 4–6, 2016.
[122] "Draft information security law of Vietnam." [Online]. Available: http://duthaoonline.quochoi.vn/DuThao/Lists/DT_DUTHAO_LUAT/View_Detail.aspx?ItemID=655&LanID=656&TabIndex=1.
[123] T. P. Thanh and H. M. Duc, "Truong Phuoc Thanh and Ha Minh Duc INFORMATION SOCIETY OF VIETNAM : CURRENT STATE AND PERSPECTIVE Thesis Degree Programme in Information Technology," no. April, 2012.
[124] B. Hoang and M. Cuong, "Current Status of Vietnamese E-commerce," 2003.
[125] D. Le, "Challenges of Internet Development in Vietnam :," no. January, pp. 16–19, 2007.
[126] L. Science, "VIET NAM REPORT (Final report of the second phase) Institute of Labour Science and Social affairs (ILSSA)," 2004.
[127] M. Musil, "VISEGRAD GROUP – AFTER 20 YEARS," pp. 429–447, 2011.
[128] S. I. Mária, "Problems and Future Possibilities of Visegrad Cooperation The Problems of Visegrad Cooperation," vol. 13, no. 2, pp. 295–304, 2014.
[129] "Visegrád history." [Online]. Available: http://www.visegradgroup.eu/about/history.
[130] V. Brazova, P. Matczak, and V. Takacs, "Regional Organization Study: Visegrad Group," no. July, 2013.
[131] L. V. Helšusová, "Existence and signification of the Visegrad Group in the perspective of its citizens," 2003.
[132] Z. Ill and K. Gapi, "V4 GOES CYBER : CHALLENGES AND OPPORTUNITIES."
[133] "Visegrad group 's aims in cooperation." [Online]. Available: http://www.visegradgroup.eu/about.
[134] "Central European Countries Security Platform." .
[135] "Guidelines on the Future Areas of Visegrad Cooperation," 2018. [Online]. Available: http://www.visegradgroup.eu/cooperation/guidelines-on-the-future-110412.

[136] "Visegrád battlegroup." [Online]. Available: https://www.globalsecurity.org/military/world/europe/visegrad.htm.
[137] "Visegrád group." [Online]. Available: http://www.visegradgroup.eu/about.
[138] A. Orosz, "The Western Balkans on the Visegrad Countries ' Agenda," 2017.
[139] "Fields of Cooperation between the Visegrad Group Countries and the Benelux," 2005. [Online]. Available: http://www.visegradgroup.eu/2005/fields-of-cooperation.
[140] "Summit Meeting between Benelux and the Visegrad Group Luxembourg (5 December 2001)." [Online]. Available: http://www.visegradgroup.eu/2001/summit-meeting-between.
[141] "Guidelines on the Future Areas of Visegrad Cooperation." [Online]. Available: http://www.visegradgroup.eu/cooperation/guidelines-on-the-future-110412.
[142] A. Jiřičková, "V4 defence cooperation," *Assoc. Int. Aff. 21st Seas. Prague Student Summit*, pp. 1–25, 2015.
[143] "New dimension of V4." [Online]. Available: http://visegradplus.org/analyse/new-dimension-v4-defense-cooperation-comparative-analysis-cybersecurity-strategies-cecsp-countries/.
[144] L. Foundations and R. E. Text, "COUNTRY : HUNGARY The National Cyber Security Strategy of Hungary was," pp. 4–6, 2013.
[145] Czech Republic - Ministry of Foreign Affairs, "Security Strategy of the Czech Republic," no. January, pp. 4–7, 2015.
[146] L. Foundations and R. E. Text, "COUNTRY : POLAND," pp. 1–2, 2013.
[147] L. Foundations and R. E. Text, "COUNTRY : SLOVAKIA," pp. 1–3, 2014.
[148] Czech Republic - Ministry of Foreign Affairs, "Security Strategy of the Czech Republic 2011," 2015.
[149] Ministry of Foreign Affairs of the Czech Republic, "The National Security Strategy of the Czech Republic 2015," 2015.
[150] Ministry of Administration and Digitisation Internal Security Agency, "Cyperspace Protection Policy of the Republic of Poland," p. 26, 2013.
[151] Hungary, "National Cyber Security Strategy of Hungary," vol. 2013, no. 1139, pp. 1–6, 2013.
[152] László Kovács, "Hungary national cybersecurity strategy," *Tallinn*, 2015.
[153] Cyberwiser.eu, "NATIONAL CYBER SECURITY STRATEGY - NIS Capacities." [Online]. Available: https://cyberwiser.eu/hungary-hu.
[154] GoS, "National Strategy for Information Security in the Slovak Republic," 2008.
[155] K. Kaska, "National Cyber Security Organisation : Slovakia," 2015.
[156] Slovak Republic, "Cyber Security Concept of the Slovak Republic," 2015.
[157] C. Leuprecht, D. B. Skillicorn, and V. E. Tait, "Beyond the Castle Model of cyber-risk and cyber-security," *Gov. Inf. Q.*, vol. 33, no. 2, pp. 250–257, 2016.
[158] European Commission, "Cybersecurity Strategy of the European Union: An Open, Safe and Secure Cyberspace - Joint Communication to the European

Parliament, the Council, the European Economic and Social Committee and the Committee of the Regions," *Eur. Comm.*, p. 20, 2013.
[159] C. Intersoft, "General Data Protection Regulation (GDPR)." [Online]. Available: https://gdpr-info.eu/.
[160] EUREKA, "Article 32 of the GDPR Law – effects and implications," 2017. [Online]. Available: https://eureka.eu.com/gdpr/article-32/.
[161] Imperva, "GDPR Article 32," 2019. [Online]. Available: https://www.imperva.com/data-security/regulation-glossary/gdpr/gdpr-article-32/.
[162] W. L. Ross and K. Rochford, "Draft NIST Special Publication 800-53 Security and Privacy Controls for Information Systems and Organizations," 2017.
[163] N. Lord, "What is NIST SP 800-53? Definition and Tips for NIST SP 800-53 Compliance," 2018. [Online]. Available: https://digitalguardian.com/blog/what-nist-sp-800-53-definition-and-tips-nist-sp-800-53-compliance.
[164] R. S. Team, "How to Use NIST Frameworks for GDPR Requirements," 2018. [Online]. Available: https://www.riskmanagementstudio.com/how-to-use-nist-frameworks-for-gdpr-requirements/.
[165] S. B. C. C. Team, "NIST Releases Fifth Revision of Special Publication 800-53," 2017. [Online]. Available: https://www.insidegovernmentcontracts.com/2017/08/nist-releases-fifth-revision-special-publication-800-53/.
[166] E. C. Organization, "Introduction of the Contractual public private partnership (cPPP)," 2019. [Online]. Available: https://ecs-org.eu/cppp.
[167] European Commission, "What is Horizon 2020?" [Online]. Available: https://ec.europa.eu/programmes/horizon2020/en/what-horizon-2020.
[168] I. E. Group, *Mid-term review of the contractual Public Private Partnerships (cPPPs) under Horizon 2020*. 2020.
[169] COM, "EUROPE 2020 A strategy for smart, sustainable and inclusive growth," *Com 2020*, p. 37, 2010.
[170] European Commission, "Digital Single Market," 2019. [Online]. Available: https://ec.europa.eu/digital-single-market/.
[171] "The Three Seas Initiative."
[172] T. K. Institute, "THE KOSCIUSZKO INSTITUTE POLICY BRIEF THE DIGITAL 3 SEAS INITIATIVE : A CALL FOR A CYBER UPGRADE OF REGIONAL COOPERATION WHITE PAPER," vol. 0.
[173] R. Sabillon, V. Cavaller, and J. Cano, "National Cyber Security Strategies: Global Trends in Cyberspace," *Int. J. Comput. Sci. Softw. Eng.*, vol. 5, no. 5, pp. 2409–4285, 2016.
[174] ENISA, *EP3R 2010-2013 Four Years of Pan-European Public Private Cooperation*, no. November. 2014.
[175] M. Małecki, "V4: 'It's good to be among friends,'" 2018. [Online]. Available: https://warsawinstitute.org/v4-good-among-friends/.
[176] J. Davis, "Slovakia's Leadership in Public Sector Cybersecurity Will Benefit the Visegrád Group and Beyond," 2017. [Online]. Available: https://researchcenter.

paloaltonetworks.com/2017/12/cso-slovakias-leadership-public-sector-cybersecurity-will-benefit-visegrad-group-beyond/.

[177] The Kosciuszko Institute, "Digital 3 seas Initiative cooperation," 2018. [Online]. Available: https://ik.org.pl/en/projects/thedigital3seasinitiative/.

[178] "Why Data Classification?," 2014.

[179] Deloitte, "Cyber regulation in Asia Pacific: How financial institutions can craft a clear strategy in a diverse region," *Deloitte Touche Tohmatsu*, 2017.

[180] "Internet users in Asia 2017." [Online]. Available: https://www.internetworldstats.com/stats3.htm#asia.

[181] "Asia Pacific Computer Emergency Response Team." [Online]. Available: https://www.apcert.org/about/index.html.

[182] N. Blackmore, "Data protection in Hong Kong : overview," *Data Prot.*, vol. 7567, pp. 1–20, 2016.

[183] DLA PIPER, "Data protection laws of Hong Kong," no. February 2019, 2015.

[184] D. Protection, "Data Protection Japan," *Nonscholar*, no. May, 2016.

[185] DLA PIPER, "Data protection laws South Korea," no. February 2019, 2015.

[186] D. Protection, "Data Protection Singapore," *Nonscholar*, no. May, 2016.

[187] D. Protection, "Data Protection Malaysia," *Nonscholar*, no. May, 2016.

[188] D. Protection, "Data Privacy in the Philippines," *Nonscholar*, no. May, 2016.

[189] Terence Lee, "'Anonymous' hackers threaten war with Singapore government," 2013. [Online]. Available: https://www.techinasia.com/youtube-anonymous-hacker-group-threatens-war-singapore-govt-video-removed-viral.

[190] T. Rsis, W. Paper, I. Studies, U. If, and R. W. Papers, "No . 263 Regional Cyber Security : Moving Towards a Resilient ASEAN Cyber Security Regime Caitríona H . Heinl S . Rajaratnam School of International Studies Singapore 09 September 2013 About RSIS," no. 263, 2013.

[191] ATKearney, "Cybersecurity in ASEAN: An Urgent Call to Action," 2018.

[192] R. O. Storey, "Gemalto warns against dangerous IT security complacency," 2009. [Online]. Available: https://www.networkworld.com/article/2273097/lan-wan/gemalto-warns-against-dangerous-it-security-complacency.html.

[193] AseanCU org, "ASEAN cyber university project." [Online]. Available: http://www.aseancu.org/pr/contents/acu/history.acu.

[194] 2015 AUN Secretariat, "ASEAN Cyber project with KOK." [Online]. Available: http://www.aunsec.org/aseankoreaacademic.php.

[195] A. Secretariat, "Joint Communique of the Third ASEAN Ministerial Meeting on Transnational Crime (AMMTC) Singapore, 11 October 2001." [Online]. Available: http://asean.org/?static_post=joint-communique-of-the-third-asean-ministerial-meeting-on-transnational-crime-ammtc-singapore-11-october-2001.

[196] "Asia Pacific Computer Emergency Response Team." [Online]. Available: http://www.apcert.org/about/structure/members.html.

[197] A. Pacific, C. Emergency, and R. Team, "Activities, Challenges & Collaboration (APCERT)," no. February, pp. 1–19, 2018.

[198] "The organization of the Islamic Cooperation – Computer Emergency Response Team (OIC-CERT)," 2018. [Online]. Available: https://www.oic-cert.org/en/fullmembers.html#.XAqhL-J7mUk.
[199] "China national cybersecurity strategy," 2016. [Online]. Available: https://chinacopyrightandmedia.wordpress.com/2016/12/27/national-cyberspace-security-strategy/.
[200] T. Dalgleish *et al.*, "China and Cyber: Attitudes, Strategies, organizations," *J. Exp. Psychol. Gen.*, vol. 136, no. 1, pp. 23–42, 2007.
[201] D. Protection, "Data Protection in China," *Nonscholar*, no. May, 2016.
[202] "International strategy of cooperation on cyberspace," 2017. [Online]. Available: http://www.chinadaily.com.cn/kindle/2017-03/02/content_28409210.htm.
[203] G. network of director Institutes, "Guiding Principles for cybersecurity oversight," *Sustain. Sci. Pract. Policy*, vol. 11, no. 1, pp. 1–5, 2015.
[204] "Hong Kong Information & Cyber Security." [Online]. Available: https://www.ogcio.gov.hk/en/our_work/information_cyber_security/collaboration/.
[205] D. P. Bastos and B. G. Giusto, *The privacy, data protection and cybersecurity law review: Brazil*, vol. 4. 2017.
[206] Office of the Government Chief Information Officer, "Practice Guide for Information Security Incident Handling [ISPG-SM02]," no. November, 2017.
[207] B. T. Population *et al.*, "CYBERWELLNESS PROFILE OF HONG KONG," vol. 000, no. December 2012, pp. 2012–2014, 2013.
[208] UNFCCC, "GovCERT. HK Annual Report 2017," *Park. Relat. Disord.*, vol. 21, no. 5, p. 430, 2017.
[209] APCERT Secretariat: JPCERT/CC, "APCERT CYBER DRILL 2018 'DATA BREACH VIA MALWARE ON IOT,'" pp. 1–6, 2018.
[210] S. Uesugi, "From e-Japan to u-Japan Japan ' s ICT Policy Movements Beginning of e-Japan."
[211] D. Director-general, "Cybersecurity Strategy in Japan," 2014.
[212] "Japanese government released new cyber security standards for government agencies," 2014. [Online]. Available: http://www.space-cyber.jp/topics/2014/0521.php.
[213] I. S. P. Council, "Common Standards of Information Security Measures for Government Agencies," 2014.
[214] S. W. Harold *et al.*, "NATIONAL SECURITY RESEARCH DIVISION U.S .– Japan Alliance Conference Strengthening Strategic Cooperation."
[215] "How Japan Is Aiming to Close the Cybersecurity Skills Gap Before Tokyo 2020." [Online]. Available: https://researchcenter.paloaltonetworks.com/2017/05/cso-japan-aiming-close-cybersecurity-skills-gap-tokyo-2020/.
[216] P. Kallender, "Japan, the Ministry of Defense and Cyber-Security: Progress and Pitfalls," *RUSI J.*, vol. 159, no. 1, pp. 94–103, 2014.
[217] A.-L. Dardenne, "Cybersecurity: the potential for Japan-India cooperation," 2018. [Online]. Available: http://theasiadialogue.com/2018/05/30/japanese-cybersecurity-and-the-potential-for-japan-india-cooperation/.

[218] I. Security and P. Council, "Information Security 2012," 2012.
[219] Asia Pacific Cybersecurity Dashboard, "COUNTRY : SOUTH KOREA," pp. 4–7, 2011.
[220] S. Korea, "National Cyber Security Masterplan South Korea." 2011.
[221] J. A. Lewis, M. A. Porrúa, A. Catalina, G. De, and A. Díaz, "Advanced Experiences in Cybersecurity Policies and Practices," no. July, 2016.
[222] A. Choi, "Korea cybersecurity," 2018. [Online]. Available: https://www.export.gov/article?id=Korea-Cyber-Security.
[223] "Korea Information Security Management System," 2002. [Online]. Available: https://aws.amazon.com/compliance/k-isms/.
[224] T. K. Park and K. I. & S. Agency, "Korean Cybersecurity Framework," 2009.
[225] ITU, "Cyberwellness Profile Republic of Korea," vol. 000, no. December 2012, pp. 2012–2014, 2015.
[226] J. Jun, S. LaFoy, and E. Sohn, *North Korea's Cyber Operations*, no. December. 2015.
[227] H. Boo, "Chapter 2 An Assessment of North Korean Cyber Threats Hyeong-wook Boo," pp. 21–36, 2013.
[228] P. M. Cronin, "North Korea cybersecurity strategy," 2014. [Online]. Available: http://english.donga.com/3/all/26/409646/1.
[229] A. Atamanov and A. Mamaev, "North Korea : How DPRK Created World ' s Most Effective Cyber Forces."
[230] A. Mansourov, "North Korea's Cyber Warfare and Challenges for the U.S.-ROK Alliance," pp. 1–14, 2014.
[231] Cyber Security Agency of Singapore, "Simgapore's Cyber Security Strategy," no. 12, p. 2015, 2011.
[232] "Singapore cybersecurity Bill." [Online]. Available: http://www.osborneclarke.com/insights/singapores-new-cyber-security-bill-10-things-you-need-to-know/.
[233] G. Rocher and J. Brown, "The Definitive Guide to Grails," *Source*, vol. 12, no. 3, p. pp. 648, 2009.
[234] K. Kwan, "National cybersecurity strategy aims to make Smart Nation safe: PM Lee."
[235] C. Vu and S. Rajaratnam, "Cyber Security in Singapore," no. December, 2016.
[236] AFSEC, "Malaysia cybersecurity," *Air Force Saf. Cent.*, no. 726630, pp. 1–4, 2000.
[237] BSA The Software Alliance, "Cybersecurity COUNTRY : MALAYSIA," *Asia-Pacific cybersecurity*, pp. 1–3, 2017.
[238] C. Malaysia, "Corporate Profile," *J. Chem. Inf. Model.*, 2013.
[239] MOSTI, "The National Cyber Security Policy," 2010.
[240] A. C. Abdul Wahab Chief Executive Officer S M, "The role of CyberSecurity Malaysia towards cyber security industry development in Malaysia," no. June, 2015.

[241] "The Philippines country profile." [Online]. Available: https://www.bbc.com/news/world-asia-15521300.

[242] "Cybersecurity threats to cost organizations in the Philippines US$3.5 billion in economic losses," 2018. [Online]. Available: https://news.microsoft.com/en-ph/2018/06/01/cybersecurity-threats-to-cost-organizations-in-the-philippines-us3-5-billion-in-economic-losses/.

[243] "Working Draft v1.13 as of National Cybersecurity Plan December 2016 (Philippines)," no. December, pp. 1–30, 2016.

[244] E. Glukhov, "Philippines National cybersecurity 2005," *J. Biol. Chem.*, vol. 280, no. 40, pp. 33960–33967, 2005.

[245] "The Philippines cybercrime policies/ strategies," 2017. [Online]. Available: https://www.coe.int/en/web/octopus/country-wiki/-/asset_publisher/hFPA5fbKjyCJ/content/philippines/pop_up?_101_INSTANCE_hFPA5fbKjyCJ_viewMode=print&_101_INSTANCE_hFPA5fbKjyCJ_languageId=it_IT.

[246] J. Gavilan, "The state of cybersecurity in the Philippines," 2016. [Online]. Available: https://www.rappler.com/newsbreak/in-depth/130883-state-cybersecurity-philippines.

[247] "Addressing Cyberspace Vulnerability: The ASEAN and the Philippines," 2017. [Online]. Available: http://www.fsi.gov.ph/addressing-cyberspace-vulnerability-the-asean-and-the-philippines/.

[248] "Terrorism in the Philippines and U.S.-Philippine security cooperation," 2017. [Online]. Available: https://www.brookings.edu/opinions/terrorism-in-the-philippines-and-u-s-philippine-security-cooperation/.

[249] "ASEAN coopertion overview." [Online]. Available: https://asean.org/asean/about-asean/overview/.

[250] "Japan: The Philippines' most reliable and important security partner," 2017. [Online]. Available: https://www.bworldonline.com/japan-philippines-reliable-important-security-partner/.

[251] "Indonesia launches Cyber Security Agency." [Online]. Available: https://www.opengovasia.com/articles/6563-indonesia-launches-cyber-security-agency-in-wake-of-growing-threat-landscape.

[252] E. T. Asia Pacific Cybersecurity Dashboard, "COUNTRY : INDONESIA," pp. 1–4, 2008.

[253] M. Ashari, T. Roberts, G. Cyber, S. Capacity, I. Brown, and I. Instutite, "The future of cybersecurity capacity in Indonesia."

[254] MCIT, "Leading e-government institutions and officials at national level Indonesia," vol. 16001, no. 17, pp. 1–4.

[255] "Indonesia cybercrime policies/strategies," 2017.

[256] S. E. E. L. Profile, "Cybercrime policies / strategies Cybercrime legislation State of cybercrime legislation," pp. 4–7, 2018.

[257] Z. A. Hasibuan, "Indonesia National Cyber Security Strategy: Security and Sovereignty in Indonesia Cyberspace," pp. 1–37, 2013.

[258] SEACOOP, "STRENGTHENING COOPERATION ON ICT RESEARCH BETWEEN E UROPE AND S OUTHEAST A SIA ICT policies , programmes and research priorities in the 10 ASEAN countries," no. June, 2010.
[259] G. Winley, C. Arjpru, and J. Wongwuttiwat, "National information technology policy in Thailand: a comparison among organizational sectors," *Electron. J. Inf. Syst. Dev. Ctries.*, vol. 31, no. 2007, pp. 1–28, 2007.
[260] P. Ateetanan, "Country Report, Thailand 2001," vol. 3.
[261] N. E. and C. T. Center, N. S. and T. D. Agency, and M. of S. and Technology, "Ict2020 - Thailand Information and Communication Technology Policy Framework (2011-2020)," no. May, 2011.
[262] "ThaiCERT information." [Online]. Available: https://www.thaicert.or.th/about-en.html.
[263] "ICT for poverty reduction in Lao." [Online]. Available: https://unchronicle.un.org/article/ict-poverty-reduction-lao-pdr.
[264] "LAOCERT." [Online]. Available: https://www.laocert.gov.la/en/Page-1-.
[265] K. Sounnalat, "ITU Cyber security Forum and Cyber Drill Cyber Security," no. December, 2013.
[266] O. Review *et al.*, "International Multilateral partnership against cyber threats," 2011.
[267] LaoCERT, "Current status on Cyber Security Organization Chart Ministry of Post and Telecommunications Minister and 2 Vice ministers," 2018.
[268] "Laos signed MOU with Lina Network corporation in using blockchain for government." [Online]. Available: http://ictnews.vn/internet/blockchain/lao-la-quoc-gia-dau-tien-o-dong-nam-a-thuc-day-xay-dung-chinh-phu-van-hanh-bang-blockchain-168741.ict.
[269] LaoCERT, "By: Khamla SOUNNALAT Acting Director General of LaoCERT."
[270] P. PHISSAMAY, "ICT policy & development in Laos," no. October, 2016.
[271] A. Telecommunity, "Telecommunication / ICT Policy and Regulation in Lao PDR," no. August, pp. 3–5, 2015.
[272] K. Soulivong and D. General, "ICT & e-Government in Laos."
[273] S. KITTIGNAVONG, "ICT Development in Lao," no. July, 2009.
[274] F. Michael, Minges; Vanessa, Gray; Lucy, "Khmer internet: Cambodia case study," 2002.
[275] S. Nguon, M. Thesis, and T. Kerikm, "Faculty of Social Sciences Cambodia ' s Effort on Cybersecurity Regulation : Policy and Human Rights ' Implications," 2017.
[276] International Data Corporation (IDC), "Executive Summary Indonesia ICT Market Landscape Study," pp. 1–10, 2016.
[277] The Royal Government of Cambodia (RGC), "Cambodian-ICT-Masterplan-2020.pdf." 2014.
[278] N. V. T. Khanh, "The critical factors affecting E-Government adoption: A Conceptual Framework in Vietnam," *Eur. J. Bus. Soc. Sci.*, vol. 2, no. 11, pp. 37–54, 2014.

[279] D. Nguyen and R. Zoltán, "THE CURRENT STATE OF INFORMATION COMMUNICATION TECHNOLOGY IN CRITICAL INFRASTRUCTURE : THE CASE OF VIETNAM," *Hadmérnök*, no. XII, pp. 173–179, 2017.

[280] S. Republic and M. O. F. Information, "SOCIALIST REPUBLIC OF VIETNAM Building e-government and applying information technology in governmental bodies ' s activities," 2008.

[281] UN, *E-Government Survey 2010*. 2010.

[282] H. P. D. Nguyen, R. Zoltán, and D. T. Binh, "THE IMPACT OF E-LEARNING TOWARDS SMALL AND MEDIUM SIZED ENTERPRISE IN VIETNAM," in *ICFE 2016 – The 3rd International Conference on Finance and Economics Ton Duc Thang University, Ho Chi Minh City, Vietnam June 15th-17th, 2016*, 2016.

[283] D. Dai, Nguyen; Binh, "The Impact of e-Commerce in Vietnamese SMEs," *Techbullion*, vol. 3, no. 2, pp. 90–95, 2017.

[284] H. Hoa, "Vietnam leads SEA in cyber-attacks," 2018. [Online]. Available: http://ven.vn/vietnam-leads-sea-in-cyber-attacks-34351.html.

[285] "Thông tin 400.000 hành khách của Vietnam Airlines có thể chứa mã độc," 2016. [Online]. Available: https://vnexpress.net/so-hoa/thong-tin-400-000-hanh-khach-cua-vietnam-airlines-co-the-chua-ma-doc-3444506.html#ctr=related_news_click.

[286] "Sân bay Nội Bài, Tân Sơn Nhất bị tin tặc tấn công," 2016. [Online]. Available: https://vnexpress.net/thoi-su/san-bay-noi-bai-tan-son-nhat-bi-tin-tac-tan-cong-3444469.html.

[287] "Ngân hàng tại VN 'lo tin tặc tấn công,'" 2016. [Online]. Available: https://www.bbc.com/vietnamese/vietnam/2016/08/160801_vietnam_banks_review_online_security.

[288] "Vietnam hit by 6,500 cyber attacks in eight months," 2018. [Online]. Available: https://english.vietnamnet.vn/fms/science-it/208055/vietnam-hit-by-6-500-cyber-attacks-in-eight-months.html.

[289] Vietnamnet, "Vietnam hit by 6,500 cyber attacks in eight months," 2018. [Online]. Available: https://english.vietnamnet.vn/fms/science-it/208055/vietnam-hit-by-6-500-cyber-attacks-in-eight-months.html.

[290] C. T. Dai, "Cybersecurity Governance Framework in Vietnam: State of Play, Progress and Future Prospects," pp. 86–98, 2017.

[291] "VNCERT & PwC Vietnam hold cyber security drill," 2018. [Online]. Available: http://www.vneconomictimes.com/article/business/vncert-pwc-vietnam-hold-cyber-security-drill.

[292] "VNCERT introduction." [Online]. Available: http://vncert.gov.vn/gioi-thieu.php.

[293] L. A. W. O. N. E-transactions, "The Law on information transaction 2005," no. 51, pp. 1–15, 2005.

[294] "The law on transaction 2006," no. 80, 2006.

[295] T. H. E. Socialist and R. Of, "The socialist republic of vietnam ____________ ______________________________," no. November, pp. 1–15, 2003.

[296] "Decree No. 85/2016/ND-CP dated July 01, 2016, on the security of information systems by classification." [Online]. Available: https://vanbanphapluat.co/decree-85-2016-nd-cp-on-the-security-of-information-systems-by-classification.

[297] "Decree No. 108/2016/NĐ-CP dated July 01, 2016, detailed regulations on provision of cybersecurity products." [Online]. Available: https://thegioiluat.vn/phap-luat/decree-no-108-2016-nd-cp-dated-july-01-2016-detailed-regulations-on-provision-of-cyber-information-security-services-and-products-681/.

[298] "Pursuant to the Constitution of the Socialist Republic of Vietnam; The National Assembly hereby promulgates the Law on Information Security.," pp. 1–25, 2015.

[299] "Vietnam cybersecurity law 2018," 2018. [Online]. Available: https://vietnam-business-law.info/blog/2018/7/30/vietnams-new-cybersecurity-law.

[300] "Vietnam, Czech firms sign MoU on cyber security cooperation," 2017. [Online]. Available: https://en.vietnamplus.vn/vietnam-czech-firms-sign-mou-on-cyber-security-cooperation/110283.vnp.

[301] "India and Vietnam sign MoU in the field of cyber security," 2017. [Online]. Available: https://www.indiainfoline.com/article/news-sector-information-technology/india-and-vietnam-sign-mou-in-the-field-of-cyber-security-117011900197_1.html.

[302] "PwC Vietnam and VNCERT form Cyber Security Partnership," 2018. .

[303] "RMIT in Vietnam signs cyber security MoU," 2018. [Online]. Available: http://www.vneconomictimes.com/article/business/rmit-signs-cyber-security-mou.

[304] "Vietnam to increase cyber security capacity," 2019. [Online]. Available: http://sggpnews.org.vn/science_technology/vietnam-to-increase-cyber-security-capacity-79926.html.

[305] "National Cyber Security Center signs deal with Kaspersky for online security," 2019. [Online]. Available: National Cyber Security Center signs deal with Kaspersky for online security.

[306] "PHÊ DUYỆT ĐỀ ÁN 'ĐÀO TẠO VÀ PHÁT TRIỂN NGUỒN NHÂN LỰC AN TOÀN, AN NINH THÔNG TIN ĐẾN NĂM 2020,'" 2014. [Online]. Available: https://thuvienphapluat.vn/van-ban/Lao-dong-Tien-luong/Quyet-dinh-99-QD-TTg-nam-2014-De-an-dao-tao-phat-trien-nguon-nhan-luc-an-ninh-thong-tin-2020-219222.aspx.

[307] "PHÊ DUYỆT CHƯƠNG TRÌNH MỤC TIÊU CÔNG NGHỆ THÔNG TIN GIAI ĐOẠN 2016 - 2020," 2018. [Online]. Available: https://thuvienphapluat.vn/van-ban/Cong-nghe-thong-tin/Quyet-dinh-153-QD-TTg-2018-phe-duyet-Chuong-trinh-muc-tieu-Cong-nghe-thong-tin-2016-2020-374242.aspx.

[308] "Vietnam Information Security Day 2018 launched," 2018. [Online]. Available: https://english.vietnamnet.vn/fms/science-it/211728/vietnam-information-security-day-2018-launched.html.

[309] "BKAV releases the 1st cyber security training program in Vietnam," 2015. [Online]. Available: https://www.alotrip.com/vietnam-news-daily/bkav-releases-cyber-security-training-program-vietnam.

[310] "ASEAN Telecommunications and IT Ministers Meeting (TELMIN) Overview," *2001*. [Online]. Available: https://asean.org/asean-economic-community/asean-telecommunications-and-it-ministers-meeting-telmin/.

[311] Coward RT and Dwyer JW., "ASEAN ICT Masterplan 2015 Completion report," *Res. Aging*, vol. 12, no. 2, pp. 158–181, 1990.

[312] ASEAN Secretariat, "The Asean ICT Masterplan 2020," no. 70, p. d, 2015.

[313] "Can ASEAN Continue to Improve Cybersecurity in the Region and Beyond?," 2018. [Online]. Available: https://www.cfr.org/blog/can-asean-continue-improve-cybersecurity-region-and-beyond.

[314] "ASEAN Cybersecurity in the Spotlight Under Singapore's Chairmanship," 2018. [Online]. Available: https://thediplomat.com/2018/05/asean-cybersecurity-in-the-spotlight-under-singapores-chairmanship/.

[315] P. Parameswaran, "What's Next for the New ASEAN-Singapore Cyber Center?," 2018. [Online]. Available: https://thediplomat.com/2018/09/whats-next-for-the-new-asean-singapore-cyber-center/.

[316] N. A. Putra, "Is ASEAN Doing Enough to Address Cybersecurity Risks?," 2018. [Online]. Available: https://thediplomat.com/2018/03/is-asean-doing-enough-to-address-cybersecurity-risks/.

[317] "CIIP Guidelines The 9th ASEAN-Japan Information Security Policy Meeting," 2016.

[318] P. Parameswaran, "Japan, Singapore Sign New Cyber Pact," 2017. [Online]. Available: https://thediplomat.com/2017/09/japan-singapore-sign-new-cyber-pact/.

[319] P. Parameswaran, "What's in the New Singapore-Germany Cyber Pact?," 2017. [Online]. Available: https://thediplomat.com/2017/07/whats-in-the-new-singapore-germany-cyber-pact/.

[320] CNA/jt(aj), "Singapore, Canada agree to boost cybersecurity cooperation," 2018. [Online]. Available: https://www.channelnewsasia.com/news/singapore/singapore-canada-cybersecurity-cooperation-mou-agreement-10930094?cid=h3_referral_inarticlelinks_24082018_cna.

[321] CNA/aa(cy), "US, Singapore to collaborate on cybersecurity for ASEAN," 2018. [Online]. Available: https://www.channelnewsasia.com/news/singapore/cybersecurity-asean-us-singapore-work-together-10936910.

[322] K. Anh, "ASEAN, U.S. release joint statement on cybersecurity cooperation," 2018. [Online]. Available: http://news.chinhphu.vn/Home/ASEAN-US-release-joint-statement-on-cybersecurity-cooperation/201811/35174.vgp.

[323] H. Baharudin, "Singapore to draw up formal Asean mechanism for cyber security," 2018. [Online]. Available: https://www.straitstimes.com/singapore/singapore-to-draw-up-formal-asean-mechanism-for-cyber-security.

[324] AKT/SH, "Delhi Declaration of the ASEAN-India Commemorative Summit to mark the 25th Anniversary of ASEAN-India Dialogue Relations," 2018. [Online]. Available: http://pib.nic.in/newsite/PrintRelease.aspx?relid=175908.

[325] A. Secretariat, "ASEAN, India to strengthen strategic partnership," 2018. [Online]. Available: https://asean.org/asean-india-to-strengthen-strategic-partnership/.

[326] "India, ASEAN Agree To Deepen Cooperation In Combating Terrorism," 2018. [Online]. Available: https://www.ndtv.com/india-news/india-asean-agree-to-deepen-cooperation-in-combating-terrorism-1804874.

[327] "Launch of H2020 Project : EU-ASEAN Cooperation on Cybersecurity Awareness," 2018. [Online]. Available: http://www.eucentre.sg/?p=15461.

[328] "Project YAKSHA to reinforce EU-ASEAN cooperation in cybersecurity," 2018.

[329] "Deepening EU security cooperation with Asian partners: Council adopts conclusions," 2018. [Online]. Available: https://www.consilium.europa.eu/en/press/press-releases/2018/05/28/deepening-eu-security-cooperation-with-asian-partners-council-adopts-conclusions/.

[330] "Asia security cooperation: EU increases engagement on security in and with Asia," 2018. [Online]. Available: https://eeas.europa.eu/headquarters/headquarters-homepage/45299/asia-security-cooperation-eu-increases-engagement-security-and-asia_en.

[331] T. Macaulay, "How the Singapore government supports the country's tech scene," 2018. [Online]. Available: https://www.cio.com/article/3299480/how-the-singapore-government-supports-the-country-s-tech-scene.html.

[332] ICTNews, "Việt Nam xếp thứ 101 trên 193 nước về khả năng đảm bảo an ninh mạng," 2017. [Online]. Available: https://ictnews.vn/cntt/bao-mat/viet-nam-xep-thu-101-tren-193-nuoc-ve-kha-nang-dam-bao-an-ninh-mang-160652.ict.

[333] Bnews, "Nhân lực ngành an toàn thông tin mạng cần cả 'lượng' và 'chất,'" 2018. [Online]. Available: https://bnews.vn/nhan-luc-nganh-an-toan-thong-tin-mang-can-ca-luong-va-chat-/77086.html.

[334] J. F. and G. L. Dong Jiang, "Sustainable Urbanization in the China-Indochinese Peninsula Economic Corridor," 2016. [Online]. Available: https://www.intechopen.com/books/sustainable-urbanization/sustainable-urbanization-in-the-china-indochinese-peninsula-economic-corridor.

[335] S. Li, Q. Wang, and J. A. Chun, "Impact assessment of climate change on rice productivity in the Indochinese Peninsula using a regional-scale crop model," *Int. J. Climatol.*, vol. 37, no. December, pp. 1147–1160, 2017.

[336] K. Janssen, "Indochina: Cambodia, Laos, and Vietnam," 2008. [Online]. Available: http://maps.unomaha.edu/Peterson/geog1000/Notes/Notes_Exam3/InDoChina.html.

[337] "Vietnam country profile," 2018. [Online]. Available: https://www.bbc.com/news/world-asia-pacific-16567315.

[338] "Laos country profile," 2018. [Online]. Available: https://www.bbc.com/news/world-asia-pacific-15351898.

[339] T. Lambert, "A BRIEF HISTORY OF CAMBODIA," 2019. [Online]. Available: http://www.localhistories.org/cambodia.html.

[340] T. Lambert, "A brief History of Vietnam," 2019. [Online]. Available: http://www.localhistories.org/viethist.html.

[341] T. Lambert, "A short History of Thailand," *2019*. [Online]. Available: http://www.localhistories.org/thailand.html.

[342] Z. Reed Smith LLP - Charmian , Aw ; Xiaoyan, "Southeast Asian nations to form regional framework for cybersecurity cooperation," 2018. [Online]. Available: https://www.lexology.com/library/detail.aspx?g=844560db-ad55-48e9-aafb-ad5328c192a9.

[343] UNDOC Regional Office for Southeast Asia and the Pacific, Thailand Institute of Justice, and United Nations Office on Drugs and Crime, "Trafficking in Persons from Cambodia, Lao PDR and Myanmar to Thailand," *United Nations Off. Drugs Crime,* no. August, 2017.

[344] ITU, "About the Child Online Protection Initiative," 2019. [Online]. Available: https://www.itu.int/en/cop/Pages/about_cop.aspx.

[345] T. Carla, Licciardello; Amanda, "Celebrating 10 years of Child Online Protection," 2018. [Online]. Available: https://news.itu.int/celebrating-10-years-child-online-protection/.

[346] S. Framework, "Statistical Framework and Indicators 2010 ITU-D," 2010.

[347] ASEAN, "ASEAN Plan of Action against Human trafficking," p. 302.

[348] ASEAN, "Asean trafficking Law 1.PDF." .

[349] M. Ismail, "ASEAN: Epicentre of human trafficking," 2018. [Online]. Available: https://theaseanpost.com/article/asean-epicentre-human-trafficking.

[350] T. globla initative A. transnational O. Crime, "ASEAN & ACTIP: Using a Regional Legal Framework to Fight a Global Crime," 2017.

[351] P. Prashanth, "Singapore Ratifies ASEAN Anti-Trafficking Pact," 2016. [Online]. Available: https://thediplomat.com/2016/01/singapore-ratifies-asean-anti-trafficking-pact/.

[352] A. Secretariat, "A Step Closer for Entry into Force of the ASEAN Convention Against Trafficking in Persons," 2017. [Online]. Available: https://asean.org/a-step-closer-for-entry-into-force-of-the-asean-convention-against-trafficking-in-persons/.

[353] C. White, "Thailand ratifies ASEAN convention against human trafficking," 2016. [Online]. Available: https://www.seafoodsource.com/news/environment-sustainability/thailand-ratifies-asean-convention-against-human-trafficking.

[354] T. Debt and S. Edt, "Lao People ' S Democratic Republic Lao People ' S Democratic Republic," no. August, pp. 235–237, 2004.

[355] L. McCallum, "One Year Later: ASEAN Anti-Trafficking Action Plan Still Dormant," 2016. [Online]. Available: https://humantraffickingcenter.org/one-year-later-asean-anti-trafficking-action-plan-still-dormant/.

[356] INTERPOL, "What is INTERPOL?," 2019. [Online]. Available: https://www.interpol.int/en/Who-we-are/What-is-INTERPOL.

[357] T. Economics, "Thailand GDP per capita 2017," 2019. [Online]. Available: https://tradingeconomics.com/thailand/gdp-per-capita.

[358] T. Economics, "Lao PDR GDP per capita 2017," 2019. [Online]. Available: https://tradingeconomics.com/laos/gdp-per-capita.

[359] T. Economics, "Cambodia GDP per capita 2017," 2019. [Online]. Available: https://tradingeconomics.com/cambodia/gdp-per-capita.

[360] T. Economics, "Vietnam GDP per capita 2017," 2019. [Online]. Available: https://tradingeconomics.com/vietnam/gdp-per-capita.

[361] ASEAN Secretariat, "Free Trade Agreements with Dialogue Partners in ASEAN," 2018. [Online]. Available: https://asean.org/asean-economic-community/free-trade-agreements-with-dialogue-partners/.

[362] V. News, "Vietnam Free Trade Agreement (EVFTA) heralds new chapter for Vietnam -EU relations," 2019. [Online]. Available: https://vietnamnews.vn/media-outreach/484633/vietnam-free-trade-agreement-evfta-heralds-new-chapter-for-vietnam-eu-relations.html#z1041pQF3K2kLivY.97.

[363] A. Salikha, "LATEST: 2018 Economies & Ranking of GDP Per Capita of Southeast Asian Countries," *2018*. [Online]. Available: https://seasia.co/2018/08/10/latest-2018-economies-ranking-of-gdp-per-capita-of-southeast-asian-countries.

[364] "Russian and China cooperation in cybersecurity." [Online]. Available: https://jsis.washington.edu/news/china-russia-cybersecurity-cooperation-working-towards-cyber-sovereignty/.

[365] Besenyő János: Low-cost attacks, unnoticable plots? Overview on the economical character of current terrorism, STRATEGIC IMPACT (ROMANIA) (ISSN: 1841-5784) (eISSN: 1824-9904) 62/2017: (Issue No. 1) pp. 83-100. (2017), www.ceeol.com/search/article-detail?id=531307

[366] János Besenyő, Krisztina Márton and Ryan Shaffer: Hospital Attacks Since 9/11: An Analysis of Terrorism Targeting Healthcare Facilities and Workers, Studies in Conflict & Terrorism

[367] Besenyő János, Gulyás Attila: The effect of the dark web on the security, Journal of Security and Sustainability Issues, 2021 Volume 11 1 pp 103-121, www.tb.lt/Leidiniai/SSI/11-1/2021-11_7.pdf

ACRONYMS AND ABBREVIATION

A4	Vietnam, Thailand, Lao PDR, and Cambodia
ACID	ASEAN CERT' Incident Drill
ACTIP	ASEAN Convention Against Trafficking in Persons
ACTIVE	Advanced Cyber Threats Response Initiative
ADSL	Asymmetric Digital Subscriber Line
ADMM	ASEAN Defense Ministers Meeting
AGA	American Gaming Association
AI	Artificial Intelligence
AIM2015	ASEAN ICT Masterplan 2015
AIM2020	ASEAN ICT Masterplan 2020
AIS	Authority of Information Security
AMMTC	ASEAN Ministerial Meeting on Transnational Crime
AMS	ASEAN Member States
ANSSI	Agence Nationale de la Sécurité des Système d' Information
APCERT	Asia Pacific Computer Emergency Response Team
APEC	Asian-Pacific Economic Cooperation
APPI	Act on the Protection of Personal Information
APT	ASEAN Plus Three
APT	Advanced Persistent Attack
ARF	ASEAN Regional Forum
ASCCE	ASEAN-Singapore Cybersecurity Center of Excellence
ASEAN	Association of Southeast Asian Nations
BKAV	Bach Khoa Antivirus
B/Ds	Bureau and departments
CamCERT	Cambodian Computer Emergency Response Team
CBMs	Confidence Building Measures
CCC	The Cyber Clean Center
CCDCOE	The NATO Cooperative Cyber Defense Centre of Excellence
CECSP	Central European Security Platform
CEE	Central and Eastern European countries
CEENet	Central and Eastern European Networking Association
CEO	Chief Executive Officer

CERT	Computer Emergency Response Team
CEFTA	Central European Free Trade Association
CFSP	The Common Foreign Security Policy
CMM	Cybersecurity Capability Maturity Model
CNII	Critical National Information Infrastructure
CICA	Conference On Interaction And Confidence Building Measures In Asia
CICC	Cybercrime Investigation and Coordination Center
CII	Critical Information Infrastructure
CIIP	Critical Information Infrastructure Protection
CIPPS	Center for International Public Policy Studies
CISSP	Certified Information Systems Security Professional
CNCERT	National Computer Network Emergency Response Technical Team/ Coordination Center of China
CoE	Council of Europe
COE	Industrial Cybersecurity Center of Excellence
COP	Child Online Protection
COSTIND	The Commission for Science, Technology and Industry for National Defense
CPPP	Contractual Public Private Partnership
CRP	Cyberspace of Republic of Poland
CSA	Cybersecurity Agency of Singapore
CSA	The Cybersecurity Agency
CSCAP	Council for Security Cooperation in the Asia Pacific
CSDP	Common Security and Defense Policy
CSF	Cybersecurity Framework
CSIRT	Computer Security Incident Response Team
CSIRT.CZ	Czech Republic Computer Security Incident Response Team
CSIRT.SK	Slovakian Computer Security Incident Response Team
CSL	Cybersecurity Law
CSM-ACE	Cybersecurity Malaysia Awards, conference and Exhibition
CSIP	Cyber Security Information Portal
CSTCB	Cyber Security And Technology Crime Bureau
CyberCSI	Cybercrime Scene Investigation
CyberGURU	Cybersecurity Professional Development
CyberSAFE	Cybersecurity Awareness For Everyone
DDoS	Distributed Denial of Service
DHS	The Department of Homeland Security
DICT	Department of Information and Communication Technology
DNS	Domain Name System
DoS	Denial of Service
EAS	East Asia Summit

EAMF	Expanded ASEAN Maritime Forum
ECSC	Educational Cyber Security Center
EeB	Energy-efficient Buildings
EMU	The European Monetary Union
ENISA	European Union Agency for Network and Information Security
E3PR	European Public-Private Partnership for Resilience
ESDP	The European Security and Defense Policy
ETDA	Electronic Transactions Development Agency
EU	European Union
EVFTA	Vietnam and EU Free Trade Area
FIRST	Forum of Incident Response Security Team
FISMA	Federal Information Security Modernization Act
FOCAC	Forum on China-Africa Cooperation
FoF	Factories of the Future
FS-ISAC	Financial Services Information Sharing and Analysis Center
FTA	Free Trade Area
GBDe	Global Business Dialogue On Electronic Commerce
GCI	Global Cybersecurity Index
GDP	Gross Domestic Product
GDPR	General Data Protection Regulation
GGE	Group of Governmental Experts
GIRO	Government Information Security Incident Response Office
GOJ	Government of Japan
GovCERT	Government Computer Emergency Response Team
GovCERT.HK	Government Computer Emergency Response Team of Hong Kong
GovCERT-Hungary	Government Computer Emergency Response Team of Hungary
HDF	Hungarian Defense Forces
HPC	High Performance Computing
HTCIA	High Technology Crime Investigation Association
HKCERT	The Hong Kong Computer Emergency Response Team Coordination Center
HKPF	Hong Kong Police Force
HKIRC	Hong Kong Internet Registration Corporation Limited
HKIS	Hong Kong Internet Service Providers Association
HKISPA	Hong Kong Internet Service Providers Association
HKIX	Hong Kong Internet Exchange
HKMA	Hong Kong Monetary Authority
HTTP	Hypertext Transfer Protocol
IC3	Internet Crime Complaint Center
ICMP	Internet Control Message Protocol
ICT	Information Communication Technology

ID	Identification
ID.CERT	Indian Computer Emergency Response Team
IDSIRT	Indian Security Incidents Response Team
IHL	Institution of Higher Learning
ILO	International Labor Organization
INCB	Israel National Cyber Bureau
INDOCHINA	Indochinese peninsula
INTERPOL	International Criminal Police Organization
IoT	Internet of Things
IPA	Information Technology Agency
IRC	Internet Relay Chat
ISP	Internet Service Provider
ISIRT	Information Security Incident Response Team
It	Information Technology
ITU	International Telecommunication Unit
ITU-IMPACT	International telecommunication Union – the International Multilateral Partnership Against Cyber Threats
ISC2	The International information systems security certification consortium
ISMP	Infocomm Security Masterplan
ISPC	Information Security Policy Council
ITAS	IT association of Slovakia
JPCERT	Japan Computer Emergency Response Team
J-CSIP	Initiative For Cyber Security Information Sharing Partnership Of Japan
KCC	Korea Communications Commission
KISA	The Korea Internet & Security Agency
K-ISMS	Korea Information Security Management System
KOICA	Korean International Cooperation Agency
KrCERT	Korean Computer Emergency Response Team
RAM	Random Access Memory
RAISS	Regional Asia Pacific Information Security Standard Forum
ROK	Republic of Korea
RMF	Risk Management Framework
R&D	Research and Development
LANIC	Lao National Internet Center (LANIC)
LaoCERT	Lao Computer Emergency Response Team
Lao PDR	Lao People Democratic Republic
METI	National Information Security Center
MIC	Ministry of Internal Affairs and Communications
MIC	Ministry of Information and Communication
MICT	Ministry of Information and Communication Technology

MIIT	Ministry of Industry and Information Technology
MilCERT	Military Computer Emergency Response Team
MilCIRC	Military Computer Incident Response Capability
MIMOS	Malaysian Institute of Microelectronic Systems
MOD	Ministry of Defense
MOSTI	The Ministry of Science, Technology and Innovation
MoU	Memorandum of Understanding
MPS	Ministry of Public Security
MPTC	Ministry of Posts and Telecommunications
MTPS	Malaysia TrustMask for Private Sector
MyCERT	Malaysia Computer Emergency Response Team
MyCSC	MyCyberSecurity Clinic
MSS	Ministry of State Security
NASA	National Aeronautics and Space Administration
NATO	North Atlantic Treaty Organization
NCERT	National Computer Emergency Response Team
NCIRT	National Cyber Incident Response Team
NCIPP	National Critical Infrastructure Protection Program
NCR	National Cyber Security Research and Development
NCSS	National Cyber-Security Strategy
NCSC	National Cyber Security Center
NCSM	National Cyber Security Masterplan
NCSP	National Cyber Security Plan
NDGDM	The National Directorate General for Disaster Management
NEAC	National Electronic Authentication Center
NECTEC	National Electronics and Computer Technology Center
NGOs	Non-Governmental Organizations
NICTER	Network Incident Analysis Center For Tactical Emergency Response
NiDA	National ICT Development Authority
NPA	National Police Agency
NPPD	National Protection and Programs Directorate
NIS	Network and Information Security
NISC	National Information Security Center
NIS directive	Directive on security of Network and Information Systems
NISER	National ICT Security and Emergency Response Center
NIST	National Institute of Standards and Technology
NITC	National Information Technology Committee
NSA	National Security Authority
NSTDA	National Science and Technology Development Agency
OCSIA	Office of Cybersecurity and Information Assurance
OECD	Organization for Economic Co-operation and Development

OFCA	Office of the Communications Authority
OGCIO	Office of the Government Chief Information Officer
OIC-CERT	Organization of Islamic cooperation - Computer Emergency Response Team
OSCE	Organization For Security And Cooperation In Europe
OSINT	Open Source Intelligence
PCPD	Privacy Commissioner For Personal Data
PDPO	Personal Data Privacy Ordinance
PIPA	Personal Information Protection Act
PPC	Personal Information Protection Commission
PwC	PricewaterhouseCoopers
SANET	Slovak Academic Network
SASIB	Slovak Association for Information Security
SASTIND	The State Administration for Science, Technology and Industry for National Defense
SingCERT	Singapore Computer Emergency Response Team
SITSA	Singapore Infocomm Technology Security Authority
SOC	Security Operations Center
SOMTC	ASEAN Senior Officials Meeting on Transnational Crime
SOP	Standard Operations Procedure
TAI	Technology Achievement Index
TELMIN	Telecommunications Ministers Meeting
ThaiCERT	Thailand Computer Emergency Response team
TOC	Transnational Organized Crimes
U-Gov	Ubiquitous government
UK	United Kingdoms
UNODC	United Nations Office on Drugs and Crime
UN	United Nations
USA	United States of America
USD	United States of America Dollar
V4	Visegrád countries
VAIP	The Vietnam Association for Information Processing
VEA	Vietnam E-commerce Association
VOIP	Voice over Internet Protocol
VNCERT	Vietnam Computer Emergency Response team
VNISA	Vietnam Information Security Association
VIA	The Vietnam Internet Association
VIF	Visegrád International Fund
VSISA	The Vietnam Software and IT Services Association
WGCCIS	Working Group on Cloud Computing Interoperability Standards
WGCSP	Working Group on Cloud Security and Privacy
WGPUCS	Working Group on Provision and Use of Cloud Services

WLAN	Wireless Local Area Network
WiMAX	Worldwide Interoperability for Microwave Access
WMD	Weapon of Mass Destruction
WTO	World Trade Organization
3SI	The Three Seas Initiative

APPENDIX

APPENDIX 1:
Security cooperation between Asian countries and the USA

Alliances in security cooperation	USA
Japan	- Began in the late 18th and early 19th century. - One of closest allies and partners. - Military, economic, and political relations. - Exchange information, technology - Support for the US in missile defense system development - Establishing an alliance coordination mechanism - Expanding in maritime security, cyberspace, and outer space.
South Korea	- Established in 1950 when the US helped South Korea found a modern state is known as the Republic of Korea. -Protecting South Korea from North Korea - South Korea 's army under US operational control - Trilateral cooperation with Japan and the US - Economic partners together - A free trade agreement signed on April 1, 2007.
Australia	- Supporting the US in airstrikes to counter Islamic State in 2014 -Helping in training purposes in order to increase the number of US marines from 250 to 2500 people. -A free trade agreement signed on 18 May 2004 and effected on 1 January 2005.
Philippines	- Established bilateral cooperation in 1951 - Giving two main military station troops (Subic Bay Naval Base and Clark Air Base) - Cooperation in maritime security, disaster response, law enforcement, cybersecurity, and non – proliferation of weapons of mass destruction

Thailand	-Established diplomatic relations in 1818 and Treaty of Amity and commerce in 1833 [28]. -After world war II, the relationship was improved in diplomatic, security, and commercial relations - In 2003, it was designated as a Major Non-NATO ally. - Bilateral in economic relations - In 2013 a historic agreement on science and technology cooperation was signed - In the same international organizations: United Nations, ASEAN Regional Forum, Asia-Pacific Economic Cooperation Forum, International Monetary Fund, World Bank, and World Trade Organization. - Partner for cooperation with the organization for security and cooperation in Europe and Organization of American States observer.
Russia	- Established in 1776 - Warm-up from under the Russian President Boris Yeltsin (1991–1999) - Diplomatic and trade cooperation

APPENDIX 2:
Security cooperation between Asian countries and China

Alliances in security cooperation	China
Russia	- Bilateral cybersecurity deal was made in May 2015 - Strategic partner of cooperation and priority in diplomacy - Good-Neighborly treaty of friendship and cooperation was signed in 2001 - Bilateral relationship was improved to a comprehensive strategic partnership of coordination in 2011 - Energy cooperation for ex: a 30-year gasoline deal was signed [364]
India	- Bilateral relationship in diplomatic and economic began in 1950 (namely Sino- Indian or Indo- China) - Have conflicts in the military like the Sino- Indian war of 1962, the Chola incident in 1967, and the Sino-Indian skirmish in 1987 [34]. - In 2008, became a trade partner and start a strategic and military relationship.
USA	- Cybersecurity deal was signed in September 2015 - Bilateral relationship as a potential adversary and economic partner - Cooperation between economy, military, cultural, people to people, and sub-national areas as well as international affairs
North Korea	- Contemporary diplomatic relations in the 1930s - 1961 treaty of friendship, cooperation, and mutual assistance was signed - Cooperation in economic and energy
Japan	-Sino-Japanese friendship and trade treaty was first signed in 1871 -Sino-Japanese peace and friendship treaty was created in 1978

APPENDIX 3: Summary of cyber attacks

Cybercrimes	Types	Characteristics	Impacts
Machine-made attack	Hacking/unauthorized access to computer system or networks	- Refer to illegal access activities via a network without authorization - Using keylogger, Trojan, spyware, etc. - Purpose for reputation, profit or challenge hackers' themselves	- Make profits for hackers
	Data diddling	Unauthorized altering to data at various points during transmission Involve bank records, credit records, school transcripts, and the like.	- Changing the integrity of data
	Web jacking	- create a fake website to trap victims	- Take personal data information (user ID, password, bank account number, and so on)
	Salami attack	- small and continuous attack-> major attack - related to bank transaction	- Effect on financial issues
	Child pornography	Photographs, videos, and audio recordings which involve a prepubescent person	Negative effects on victims as injury and pain, sexually transmitted diseases, sleeplessness, depression. Sexual harassment or sexual assault and sexual crime Marital dissatisfaction, losing emotion with a spouse, or even divorce
	Spoofing and phishing	- Pretend another person to make a phone call or send emails to take sensitive data from victims	- Take personal data information (password, credit card number, etc.)

Cybercrimes	Types	Characteristics	Impacts
Man-made attack	Money laundering	-Make illegal money into legal money via legitimate financial institutions	- Financial loss
	Fraud and financial crimes	- Related to spam, auction fraud, credit card fraud, and overpayment fraud. - Money laundering	- Effect on financial issues
	Online gambling	- Refer to money transactions, money laundering.	- Lost track of time, decrease the perception of the value of cash, loss of control, legal problems, and financial ruin. - Loss of career
	Data alteration or theft	- Illegal changes as school records, bank records, and so on - Steal data information	- Change the integrity of data
	Email bombing	- Denial of service - Using zip bomb, email with Trojan, virus.	- Interrupt the system, services of network - Damage physical hardware
	Cyberbullying	- Change image, send threatening messages to victims. - Tarnish the good reputation of someone	- Effect on emotions: lost confidence, feel embarrassed and afraid of meeting people
	Steganography	- Hide secret text behind other objects as text, image, and the like	- Hard to recognize original objects.
	Computer vandalism	- Sabotage computer data, hardware.	- Damage computer hardware and data.
Both machine-made attack and man-made attack	Hacking / unauthorized access to a computer system or networks.		
	Spoofing and phishing		
	Email bombing		

APPENDIX 4:
Comparison about cyber-crime and cyber-warfare

Types	Cybercrime	Cyber-warfare
Definition	A crime which involves computer technology to access sensitive data, malicious purposes, illegal activities. Two types of cybercrimes: computer as a target of the attack, computer as a means to attack.	- An act which involves offensive and defensive activities.
Tools	Computers, malicious codes to make viruses, Trojan, malware, and so on	-Weapons combine with high-tech tools
Impact	- Strong influences on e-commerce such as integrity, authentication, availability, and authorization privacy during a business transaction. - Cause of financial damage and monetary losses. - Influence on the online and offline world. - Negative effects on business of both small and big companies - Major effects on piracy of the entertainment, music, and software industries - Spend a lot of money for building a security system	- Effect on politics, national's stability, and citizen's life. - Damage critical infrastructure (electricity power grid, water supply, transportation, control system, and so on) of a country. - Major effects on political and military communications remotely from anywhere in the world - Corrupt weapons of the enemy - Effects on health, security, or the economy, functions of government, and social wellbeing of the population

www.ingramcontent.com/pod-product-compliance
Lightning Source LLC
Chambersburg PA
CBHW071341230926
51846CB00002B/407
* 9 7 8 2 1 4 0 2 6 5 2 9 7 *